H. Gillan. Sept 1953

Sources of Mathematical Discovery

Sources of Mathematical Discovery

Lorraine Mottershead

BASIL BLACKWELL · OXFORD

First published in Australia by John Wiley and Sons Australasia Pty Ltd

First published in Great Britain by Basil Blackwell, Oxford, 1978

Cover design/Chris Belson

British Library Cataloguing in Publication Data

Mottershead, Lorraine
Sources of mathematical discovery.
1. Mathematics – Problems, exercises etc.
I. Title
510'.76 QA43
ISBN 0–631–10221–3

Printed in Hong Kong
Reprinted in Hong Kong 1981, 1983, 1984, 1985

Acknowledgments

The author and publisher gratefully acknowledge the kind permission of the following copyright holders to reproduce the material in this book.

p. 69, photograph of a pine cone by courtesy of Robert Hardie; p. 71, photograph of a nautilus shell by courtesy of Keith Gillett, FRPS, FRMS; p. 13, "Symmetry Drawing" and "Sky and Water" and "Reptiles" by courtesy of the Escher Foundation, and Haags Gemeentemuseum, The Hague; p. 130, "Study of Facial Proportions" and p. 131, "Study of Human Proportions" by courtesy of the Academy of Venice; p. 132, photograph of The Parthenon by courtesy of G. R. Sims; p. 137, the Whirlpool Galaxy, M101 by courtesy of Mt Stromlo Observatory; p. 162,, "Falsè Perspective" by courtesy of the Mansell Collection; p. 163, "Relativity", p. 164, "Waterfall", p. 165, "Belvedere", and p. 166, "Ascending and Descending" by courtesy of the Escher Foundation, Haags Gemeentemuseum, The Hague; pp. 171 and 173 Spiral and Humming Bird by courtesy of Eric Rawlings at Datamatic and the computer graphics on pages 170, 171, 172 and 173 by courtesy of Dr Herbison Evans of Sydney University's Computing Centre.

I am particularly indebted to Miss P. Playford for her interest, advice and encouragement in the preparation of this book, and special thanks must go to Ms Jane Arms for her untiring effort and unfailing co-operation in editing *Sources of Mathematical Discovery*. I would also like to thank Andrew Clarke for his work on Polyominoes and Valda Brook for drawing the illustrations.

Further Reading

Many books and articles in magazines contain excellent sections in the field of recreational mathematics and mathematical applications. The following are some I have found useful in connection with topics dealt with in *Sources of Mathematical Discovery*.

Stephen Barr, **Second Miscellany of Puzzles**. McGraw-Hill, New York, 1969.

Martin Gardner, **Further Mathematical Games**. George Allen and Unwin, London, 1970.

Martin Gardner's New Mathematical Diversions from the Scientific American. George Allen and Unwin, London, 1966.

Martin Gardner's Sixth Book of Mathematical Games from the Scientific American. W. H. Freeman, San Francisco, 1971.

Harold R. Jacobs, **Mathematics – a Human Endeavour**. W. H. Freeman, San Francisco, 1970.

Donovan A. Johnson, **Curves**. John Murray, London, 1966.

Boris A. Kordemsky, **The Moscow Puzzles (359 Mathematical Recreations)**. Charles Scribner's Sons, New York, 1972.

Charles F. Linn, **Puzzles, Patterns and Pastimes from the World of Mathematics**. Doubleday, New York, 1969.

Joseph S. Madachy, **Mathematics on Vacation**. Charles Scribner's Sons, New York, 1975.

John A. Pemberton, **Modern Geometry**. Cassell, London, 1968.

Anthony Ravielli, **Adventure with Shapes**. Phoenix House, London, 1960.

James T. Rogers, **The Story of Mathematics**. Brockhampton Press, Leicester, U.K., 1968.

William C. Vergara, **Mathematics in Everyday Things**. Harper and Row, New York, 1959.

Contents

Introduction

During the past few years many teachers have become aware of the tremendous need for motivation topics in mathematics which give the pupils an appreciation of and insight into this vast subject and help them to adopt a more positive and experimental attitude to what was formerly a purely academic subject.

Teaching techniques have to be adapted to the needs of the individual class, but where possible practical, creative work should be given to each pupil so that the spontaneous discovery of facts is satisfying and lasting. To achieve this end numerous "experiments" have been included. The onus is therefore on the teacher to make adequate preparations (usually with simple classroom materials) and gain a clear understanding of the topic.

My goal in writing this resource book was to present a variety of topics *in depth,* catering for the average student as well as the very able student who should find a challenge in most topics. It follows, therefore, that there is no obligation on all pupils to investigate all topics to the same extent; some will prove too difficult.

I have arranged the book to contain a series of different approaches to add variety and stimulate thought. For instance, there are sections on historical developments, applications of various branches of mathematics and integration of subjects, some purely for interest's sake but all, I believe, topics that students should study in their courses.

The answers to selected exercises, those marked with an asterisk in the text, are given at the back of the book.

Library research and assignments have also been included. If any comparative assessment is required or desired it could be based on the creativity of the student's approach to the content and presentation. Selected projects displayed on notice boards could be used as the basis for discussion by the class as well as creating an interesting "atmosphere".

Where possible students could prepare and present small lectures on special topics thus extending the more capable students and allowing them to gain a sense of fulfilment and the acquisition of excellence.

The conclusions that should follow from this kind of study are that mathematicians are people, that the discoveries they have made and their contributions to the subject have arisen in many cases from their interest, their curiosity and their perseverance. Hopefully pupils will absorb some of these qualities and find that mathematics has relevance to many aspects of modern living, and also that principles evolved with no apparent relation to everyday life sometimes crop up in the light of research into other fields.

Above all, it is hoped that pupils and teachers will derive enjoyment from the material in this book, because from enjoyment spring both interest and efficiency.

Finally I would like to thank my friends and colleagues for all their interest and encouragement.

Sydney, 1976. *L. J. Mottershead.*

Unit 1 A Hint of Magic

Magic Squares

Historically, the first magic square is supposed to have been marked on the back of a divine tortoise that appeared before Emperor Yu (about 2200 B.C.) when he was standing on the bank of the Yellow River. An early lo-shu (magic square) looked like this:

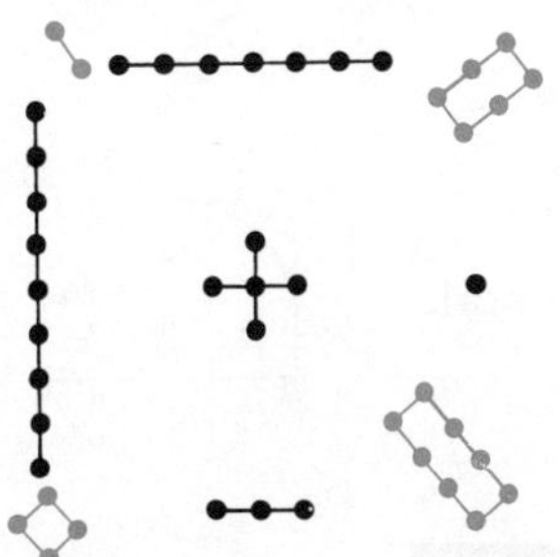

Note the rectangular and square numbers.
The **blue** circles depicted even (feminine) numbers or yin; the black circles, odd (masculine) numbers or yang.
The centre 5 represented the earth around which were the 4 elements, evenly balanced – 4 and 9 symbolising metal, 2 and 7 fire, 1 and 6 water, 3 and 8 wood.

In the Middle Ages magic squares were considered to be prophylactics (protections) against the Plague!

In the 16th century, the Italian, Cardan made an intensive study of the properties of magic squares and in the following century they were extensively studied by several leading Japanese mathematicians.

During this century they have been used as amulets in India, as well as being found in oriental fortune bowls and medicine cups. Even today, they are widespread in Tibet (appearing in the "Wheel of Life") and in other countries, such as Malaysia, that have close connections with China and India.

A well known magic square was created by the German artist Albrecht Dürer in 1514. It appears in his work entitled "Melancholia", which depicts a number of scientific instruments!

16	3	2	13
5	10	11	8
9	6	7	12
4	15	14	1

Note the date, 1514, in the centre of the bottom row.
The integers 1, 2, . . . 16 are used once only.
The sum of all the rows, columns and diagonals, as well as the sum of the four central squares, is 34.

- What is the sum of the four corner squares?
- The four 2 × 2 squares which make up each quarter of the large square add to what number?
- In each row one pair of adjacent numbers adds to 15, the other to 19. Check this.
- Add the squares of the numbers in each row:

$$16^2 + 3^2 + 2^2 + 13^2 \qquad 5^2 + 10^2 + 11^2 + 8^2$$
$$4^2 + 15^2 + 14^2 + 1^2 \qquad 9^2 + 6^2 + 7^2 + 12^2$$

- Do the same with the columns. Is the pattern similar?

16	3	2	13
5	10	11	8
9	6	7	12
4	15	14	1

In this "broken" square, add the numbers on the opposite sides together.
What is the sum in each case?

This is the centre of an ancient Tibetan seal.
All the integers from 1 to 9 are used.

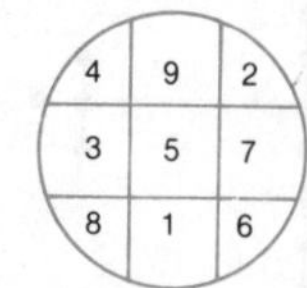

A magic square is a matrix of n^2 integers arranged in n rows and n columns in such a way that the sum of the numbers in each row, column and principal diagonal is "magic", that is, the magic constant is equal to $\frac{n}{2}(n^2 + 1)$. Use this formula to check the preceding examples.

There is a connection between magic constants and triangular numbers. For example 1, 3, 6, 10, 15 and 21 are triangular numbers and are formed by adding consecutive integers, those being 1, 1 + 2, 1 + 2 + 3, 1 + 2 + 3 + 4, 1 + 2 + 3 + 4 + 5 and 1 + 2 + 3 + 4 + 5 + 6.

The sum of the first six integers is the sixth triangular number!

In a 3 × 3 magic square formed with the integers 1 to 9, the sum of all the numbers is equal to the ninth triangular number, using the principle just described. This can be calculated as $\frac{9 \times 10}{2} = 45$. As the numbers are spread over three rows, the sum of the numbers in any one row must be one third of this total, or 15, the magic constant. There are eight ways that the digits 1 to 9 can be arranged in an order 3 array.

The numbers in a magic square, with addition, besides having the above characteristics, form sequences called arithmetic progressions.

8	1	6
3	5	7
4	9	2

Diagram 1

In this diagram, the progressions are 1, 5, 9; 3, 5, 7; 2, 5, 8 and 4, 5, 6: their common differences are 4, 2, 3 and 1 respectively.

- What is the average (mean) of the diagonals?
- What is the sum of the means of both diagonals?

- What are the four progressions in Diagram 2 and their common differences?

15	2	13
8	10	12
7	18	5

Diagram 2

These two diagrams are said to be of order 3 because they have three squares or cells or each side.

Sequences in magic squares can be used to form patterns. For instance, in Diagram 3, there is a dot marked in the centre of each square with lines through the four sets which correspond to the sequences mentioned above.

How would you describe this pattern?

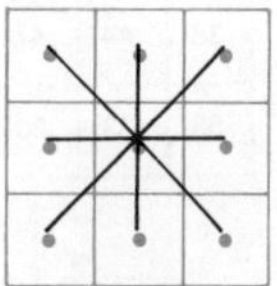

Diagram 3

EXERCISES

1. What patterns do you make by joining the dots representing the following sequences in Diagram 1?
   a. 1, 2, 3; 4, 5, 6; 7, 8, 9; [*common difference 1*].
   b. 1, 3, 5; 3, 5, 7; 5, 7, 9; [*common difference 2*].
   c. 1, 4, 7; 2, 5, 8; 3, 6, 9; [*common difference 3*].
   d. 2, 4, 6, 8. [*common difference 2*].
2. What progressions in Diagram 2 will give the same patterns as those in 1a and 1c? What are their common differences?
3. Form patterns by joining the dots corresponding to the numbers, taken in increasing order of magnitude, in Diagrams 1 and 2. (The two patterns are not alike, but all magic squares of nine cells belong to one or other of these two patterns.)

4.

15	20	13
14	16	18
19	12	17

Basic squares can be adapted by:
- changing the order of the numbers;
- adding a constant throughout, therefore making the total larger.

Check these facts using the given diagram.

5.

1	15	14	4
12	6	7	9
8	10	11	5
13	3	2	16

Work out this pattern, by joining consecutive numbers.

6.

17	59	56	26
50	32	35	41
38	44	47	29
53	23	20	62

Write down the integers of this square in increasing order.
What is the common difference between the consecutive integers?
What is the magic constant?

There are 880 known magic squares (excluding rotations and reflections) of order 4 and over 68,000 of order 5.

7. Compare the arrangement of integers in this magic square with that in question 5. Is there any similarity? Look at the pattern and the arrangement of numbers in this diagram.

16	3	2	13
5	10	11	8
9	6	7	12
4	15	14	1

8.

25	32	9	16	23
31	13	15	22	24
12	14	21	28	30
18	20	27	29	11
19	26	33	10	17

Calculate the total of each diagonal.

9. Calculate the missing digits in both the rows and columns.

21		23	18
25	16	15	28
	12		32
9	34	35	

Diabolic Magic Squares

If a magic square of orders 4, 5, 7 or 8 has a constant sum along its rows, columns, main and broken diagonals it is said to be diabolic.
The six "broken" diagonals in these squares are shown.

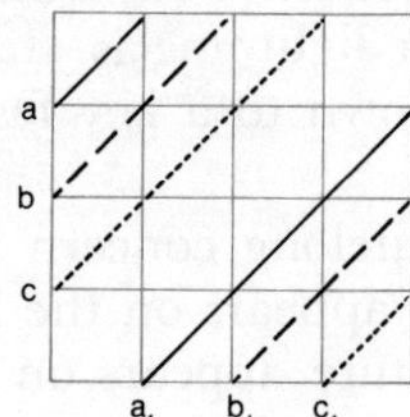

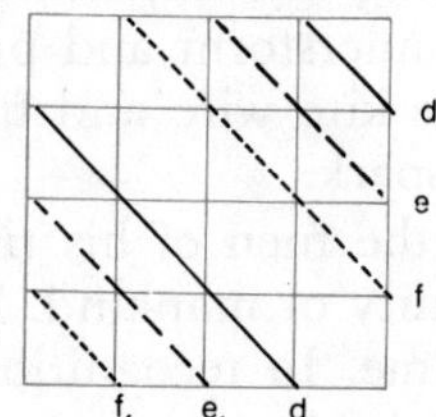

A square of order 5 has eight broken diagonals and the number increases by two with each increase in order.

What is the magic constant in this square, along:
- any column?
- any diagonal?
- the broken diagonal marked?

The twenty sets should all have the same answer.

1	8	15	17	24
20	22	4	6	13
9	11	18	25	2
23	5	7	14	16
12	19	21	3	10

From the 12th century we have an example of a **diabolic** ("devilish") square. Not only do the rows, columns and diagonals add to 34, but even the "broken" diagonals such as 12, 8, 5 and 9 add to 34. They are also called **panmagic squares.**
Study the arrangement of integers closely.

How many 34s can you find in any given square?

4	9	6	15	4	9	6
14	7	12	1	14	7	12
11	2	13	8	11	2	13
5	16	3	10	5	16	3
4	9	6	15	4	9	6
14	7	12	1	14	7	12
11	2	13	8	11	2	13

The American statesman, scientist, philosopher, author and publisher, Benjamin Franklin (1706-1790), created a magic square full of interesting features.

He was born in Massachusetts and was the fifteenth child and youngest son in a family of seventeen children. In a very full life he investigated the physics of kite flying, he invented a stove, bifocal glasses, he founded hospitals, libraries and various postal systems, and was the signer of the Declaration of Independence. He also favoured daylight saving! He worked on street lighting systems, a description of lead poisoning and experiments in electricity. In 1752, he flew a homemade kite in a thunderstorm and proved that **lightning** is **electricity.** A bolt of lightning struck the kite wire and travelled down to a key fastened at the end, where it caused a spark.

Franklin led all the men of his time in a lifelong concern for the happiness, well-being and dignity of mankind. His name appears on the list of the greatest Americans of all time. In recognition, his picture appears on some stamps and money of the United States.

He also charted the movement of the Gulf Stream in the Atlantic Ocean, recording its temperatures, speed and depth.

52	61	4	13	20	29	36	45
14	3	62	51	46	35	30	19
53	60	5	12	21	28	31	44
11	6	59	54	43	38	27	22
55	58	7	10	23	26	39	42
9	8	57	56	41	40	25	24
50	63	2	15	18	31	34	47
16	1	64	49	48	33	32	17

a. Calculate the sum of the horizontal rows, vertical columns and diagonals.
b. What is the sum of the integers within each quarter?
c. What is the total of the boxes, up four and down four? (dotted lines)
d. Calculate the sum of the four corners plus the four middle boxes.
e. Find the sum of any four-box sub-square.
f. Work out the sum of any four boxes equidistant from the square's centre.

Franklin also composed a 16 × 16 magic square with equally remarkable properties.

11. This figure has a magic constant of 150 along any row, column, main or broken diagonal.

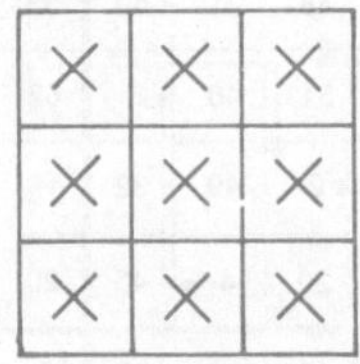

1	42	29	7	36	35
48	9	20	44	13	16
5	38	33	3	40	31
43	14	15	49	8	21
6	37	34	2	41	30
47	10	19	45	12	17

If this pattern is used with this magic square, the sum of the numbers under the crosses is equal to twenty-five times the number of crosses in the pattern.

Check this by adding the nine squares in the top left hand corner.

Try the following patterns anywhere in the large square.

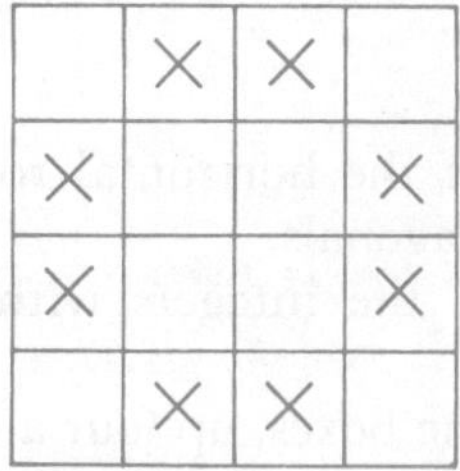

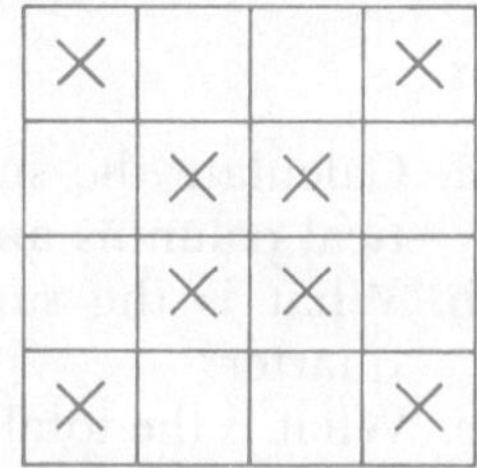

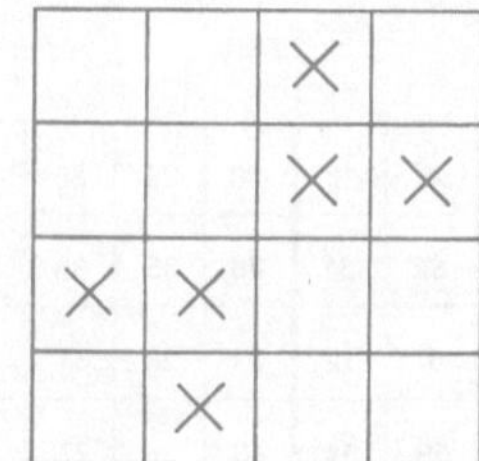

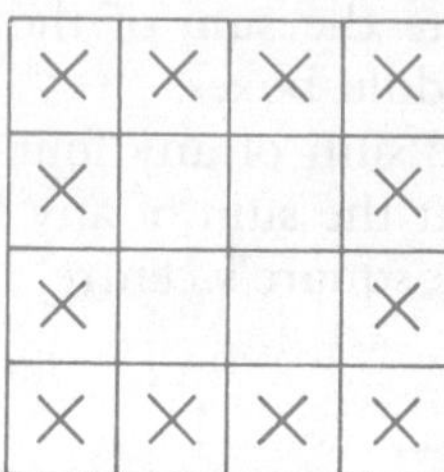

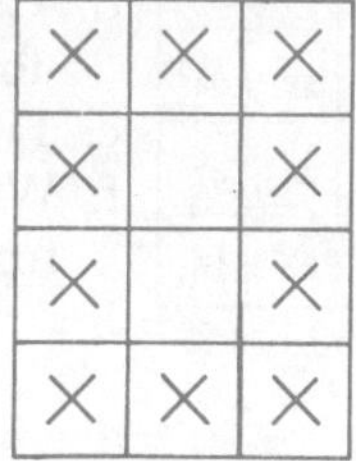

A Square for a Knight

In the 18th century Leonhard Euler (pronounced Oiler) made up this fascinating square, with rows, columns and diagonals totalling 260 and all half-way totals 130.

Even more intriguing is that a chess knight, starting its L-shaped moves (blue lines) from square 1, can "hit" all 64 numerals in order. Try it.

1	48	31	50	33	16	63	18
30	51	46	3	62	19	14	35
47	2	49	32	15	34	17	64
52	29	4	45	20	61	36	13
5	44	25	56	9	40	21	60
28	53	8	41	24	57	12	37
43	6	55	26	39	10	59	22
54	27	42	7	58	23	38	11

How To Construct Magic Squares

To construct a 3 × 3 square.

7 + 1		7 + 4
	7	
7 − 4		7 − 1

Choose any two numbers which are not in the ratio 1 : 2 or 2 : 1, then choose a third number not less than 5, which is greater than the sum of the first pair. (For example, the numbers could be 1, 4 and 7.)

Put the third number in the central cell and fill the two diagonals with numbers which form arithmetic progressions with the common differences being the original pair of numbers. Then fill in the remaining cells, remembering that the rows, columns and diagonals must all have the same total.

a − x	a + x − y	a + y
a + x + y	a	a − x − y
a − y	a − x + y	a + x

Algebraically, if x y and a are a mixture of odd and even numbers then the numbers in opposite corners will be either both odd or both even.

A 4 × 4 square.

Place the integers in the arrangement given here.

16th	3rd	2nd	13th
5th	10th	11th	8th
9th	6th	7th	12th
4th	15th	14th	1st

A 5 × 5 arrangement.

11	18	25	2	9
10	12	19	21	3
4	6	13	20	22
23	5	7	14	16
17	24	1	8	15

Use the formula $\frac{n}{2}(n^2 + 1)$ to work out the sum in this magic square, then check by addition.

Steps for construction:

- Put 1 in the centre of the bottom row.
- The next number is placed in the cell at the top of the next column on the right.

Fill the cells in a downward diagonal line, from left to right and upon filling a cell in the last right hand column, place the next number in the cell at the extreme left of the next lower row; if the next cell on the left to right downward diagonal is filled, place the next number in the cell immediately above the last one filled. After filling the cell in the lower right hand corner, place the next number in the cell immediately above it.

Variations

Is this a magic square?
What is its constant?
Add the four centre squares.
Find the sum of the four corner squares.

96	11	89	68
88	69	91	16
61	86	18	99
19	98	66	81

There are at least forty-eight different ways in which this total may be obtained. Discover some of your own. Turn the page upside down and see if the square is still magic.

x	5	x-2
7	9	11
8	x +1	x - 6

These are magic squares.
What numbers replace the pronumeral x in each diagram?

0	1	
		x
2		

A Domino Magic Square

Arrange $\frac{4}{4}$, $\frac{4}{5}$, $\frac{4}{6}$, $\frac{5}{5}$, $\frac{5}{6}$, $\frac{3}{4}$, $\frac{3}{6}$, and $\frac{6}{6}$ so that each row, column and diagonal will have the same total. (*Hint* : 19.)

A Bordered Magic Square

2	11	12	13	77	78	79	81	16
6	18	27	26	61	62	65	28	76
7	59	30	35	51	53	36	23	75
8	58	32	38	45	40	50	24	74
73	57	49	43	41	39	33	25	9
72	22	48	42	37	44	34	60	10
68	19	46	47	31	29	52	63	14
67	54	55	56	21	20	17	64	15
66	71	70	69	5	4	3	1	80

This is magic overall and each square formed by deleting a border is also magic. Check this with a calculator.

Here is a simple magic square, incorporating negative numbers.

1	-4	3
2	0	-2
-3	4	-1

Antimagic Squares have the rows, columns and diagonals adding to different totals. This is a simple one.

	9	8	7	24
	2	1	6	9
	3	4	5	12
11	14	13	18	15

A Composite Magic Square

	11	18	13		74	81	76		29	36	31	
	16	14	12		79	77	75		34	32	30	
	15	10	17		78	73	80		33	28	35	
	56	63	58		38	45	40		20	27	22	
	61	59	57		43	41	39		25	23	21	
	60	55	62		42	37	44		24	19	26	
	47	54	49		2	9	4		65	72	67	
	52	50	48		7	5	3		70	68	66	
	51	46	53		6	1	8		69	64	71	

This is similarly magic overall and simultaneously magic for each of the smaller squares. Check this carefully.

Multiplying Magic Squares

What is the constant in each square?

12	1	18
9	6	4
2	36	3

Each may be modified by changing the order of the numbers or by multiplying each number by a constant number.

36	1	48
16	12	9
3	144	4

1	12	10
15	2	4

This multiplication square makes 120 when calculated horizontally and vertically.

Complete the last row.

A Latin Square

1	2	3
2	3	1
3	1	2

This is similar to a magic square but here each row and column has the same numbers, but no number is repeated in any row or column. Are the totals of the diagonals equal?

Place 1, 2, 3 and 4 in this square so that each row column and diagonal has all four digits.

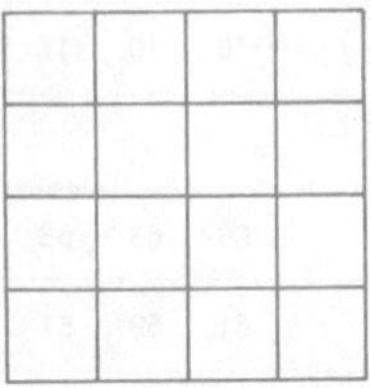

a	b	c	d
b	a	d	c
c	d	a	b
d	c	b	a

The rows and columns of this order 4 Latin square all contain the letters *a, b, c* and *d* once only. Look carefully at the arrangement. What could you say about it?

This order 5 square uses the digits 0 to 4. More complex squares called **Graeco-Latin** are used widely for experiments in agriculture, biology, medicine, sociology and even marketing.

0	1	2	3	4
3	4	0	1	2
1	2	3	4	0
4	0	1	2	3
2	3	4	0	1

A Prime Square

Is it possible to construct a magic square solely of different primes? The answer is yes.

The sum here, in all directions, is 111 (the lowest known constant for a prime square of order 3).

67	1	43
13	37	61
31	73	7

In 1913 a magic square consisting of consecutive odd primes (that is, omitting 2) was developed by J. N. Muncey. The smallest possible square of this type is one of order 12 that starts with 1 and contains the first 144 consecutive odd primes! Its rows, columns and diagonals add to 4 514.

It's a Fact . . .

The nineteen digit number 1 111 111 111 111 111 111 is prime.

Other Magic Geometric Figures

Magic Stars

A simple pentagram or five-sided star can be used to form a magic arrangement of numbers.

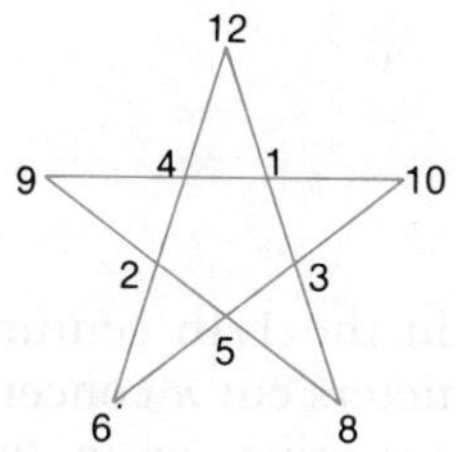

What is the sum of each straight line?
The integers 1 to 12, with the omission of 7 and 11, are used here.
It is impossible to form this type of star with only the integers 1 to 10.

In these figures, the vertices or points, are numbered 1 to 12 in such a way that the sum of the numbers along any straight line is 26 and the sum around the six points is the same.

2
1 — 12 — 6 — 7
9 10
3 — 11 — 4 — 8
5

2
4 — 11 — 5 — 6
12 9
1 — 7 — 8 — 10
3

2
6 — 5 — 12 — 3
9 11
10 — 7 — 8 — 1
4

A variation may be obtained with an eight-pointed "star". What is the sum along each line?

2
10 — 12 11 — 1
6 16
14 5
3 9
15 — 4 7 — 8
13

Magic Triangles

The integers 1 to 9 are used in the construction. Obviously, any number that is not at a vertex may be interchanged with its neighbour without affecting the sum.

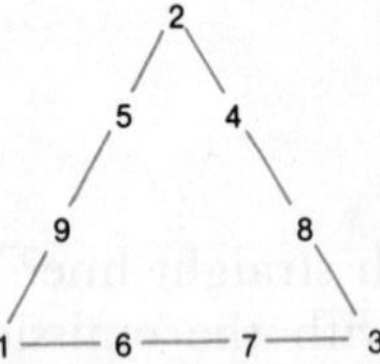

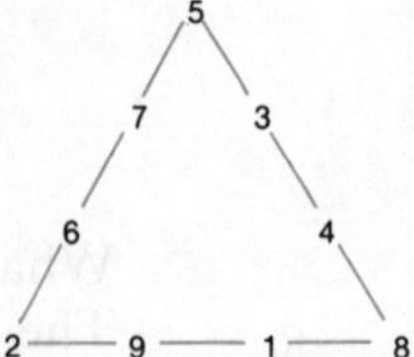

Magic Circles

Many Japanese mathematicians who studied magic squares in the 17th century also gave considerable time to magic circles. Generally, *n* diameters cut *n* concentric circles and the numbers are placed at the points of intersection, or in the annulus areas. As well, intersecting, overlapping circles can be constructed so that the points of intersection are numbered magically.

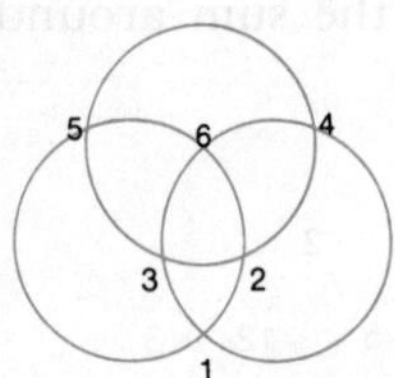

The sum of the integers on any given circle is equal to the sum of the integers on any other.

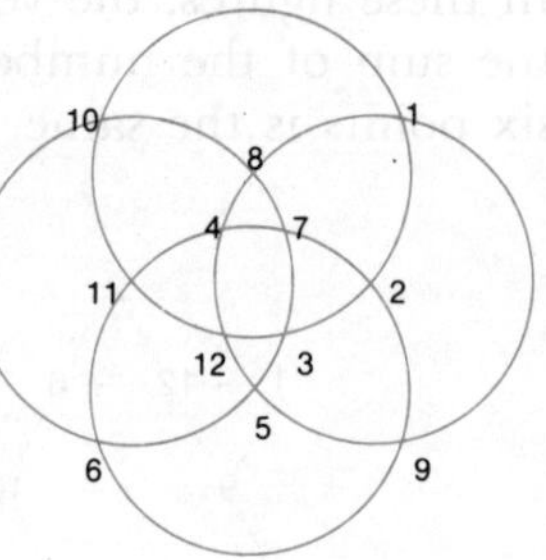

What is the sum around each circle in these diagrams?

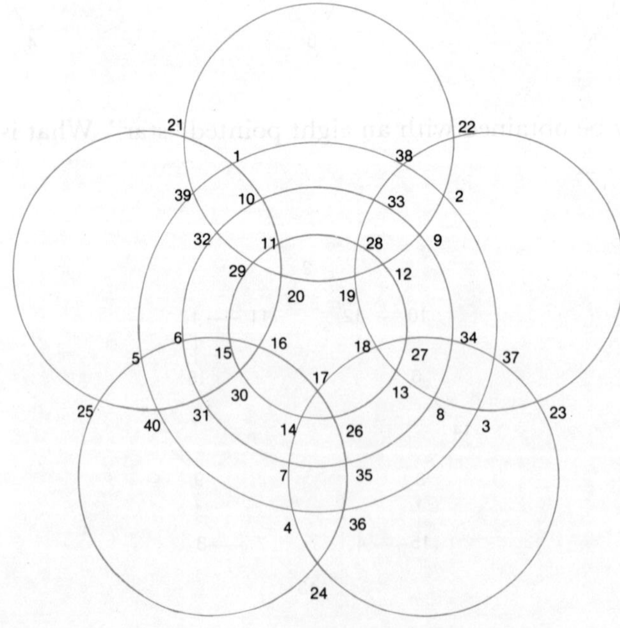

Diagram 1.

Diagram 2.

The sum of the numbers on any one diameter or circle is constant.
In Diagram 1 the integers 1 to 18 are used, being placed on the diameters in pairs so that the sum of the two numbers in corresponding positions is 19.

In Diagram 2, using any integers, we can ensure that the sum of the integers on any circle or diameter will be, say, 99, by making the sum of the two numbers in corresponding positions on any one diameter equal to 33.

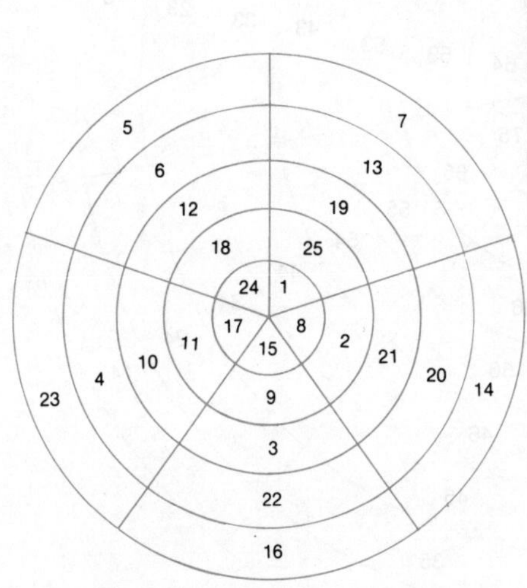

A Magic Circle of Order 5
What is the total of each sector?
What is the sum of each annulus (ring)?
What is the total of each spiral, formed by moving outwards and anticlockwise? (e.g. 1 + 18 + 10 . . .)

CONSTRUCTION

1. Draw a circle containing an odd number of sectors (greater than three) and the same number of rings.
2. Place the number 1 in any cell.
3. Put consecutive numbers in cells according to the following rules:

   a. Move outwards and clockwise.

   b. From an outermost cell move to the innermost in the next sector.

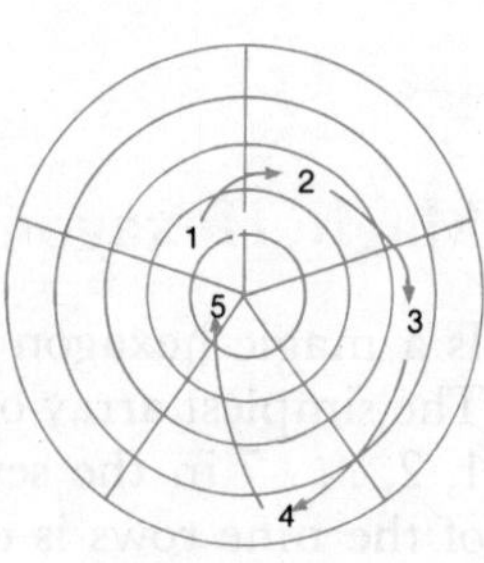

c. If the next cell is occupied, keep to the same sector and move one cell inwards.
d. From the innermost, if the next cell is occupied, move to the outermost in the same sector.

Check these instructions in the magic circle of order 9. Notice the patterns formed, as well.

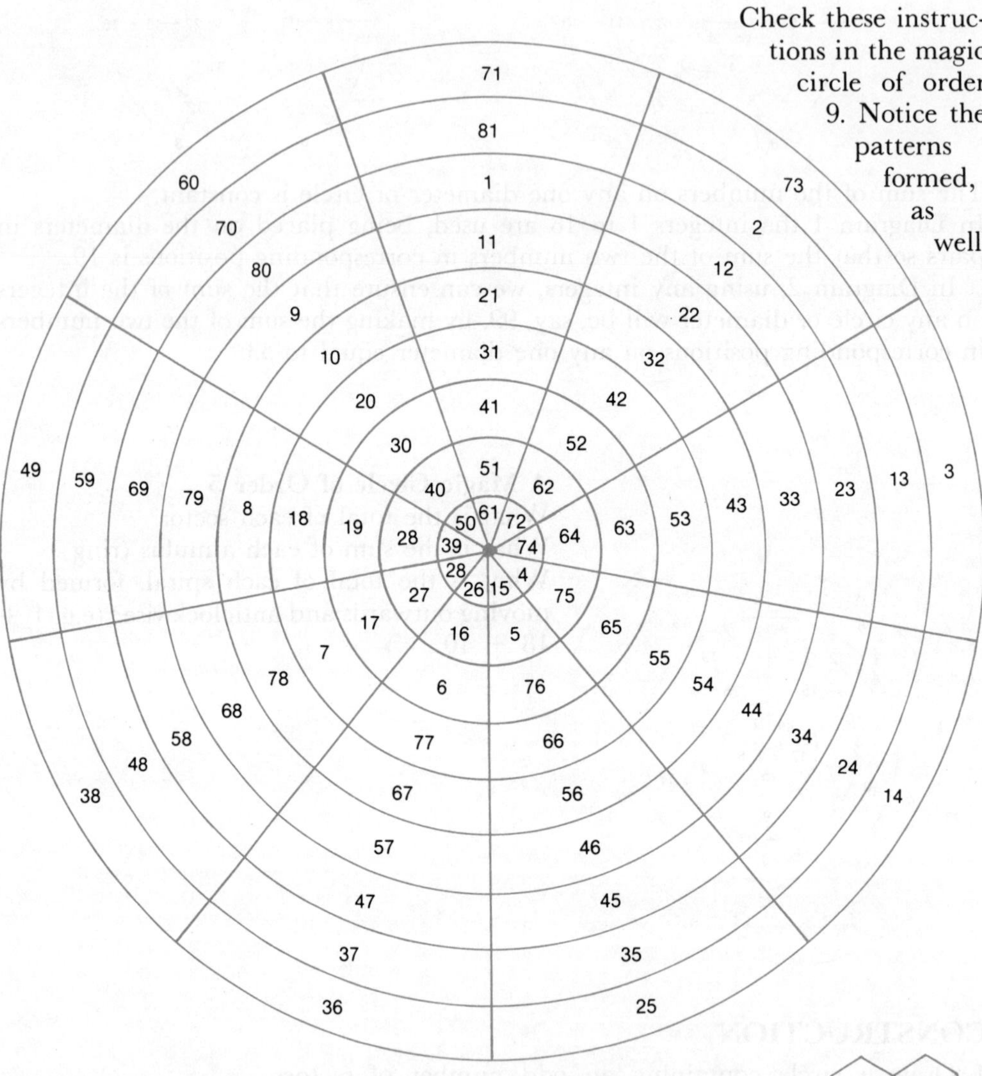

✱ Magic Hexagons

Is a magic hexagon possible?
The simplest array of cells is shown. Can you put the digits 1, 2, . . . 7 in the seven hexagons so that the sum of each of the nine rows is constant?

Order 2 hexagon

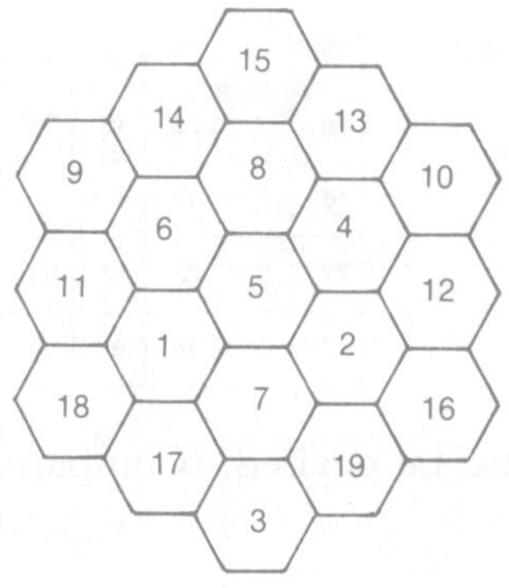

The next largest arrangement (order 3) comprises nineteen cells. The sum of the numbers is 190 which is divisible by 5, the number of parallel rows in one direction. Thus the magic constant is 38. There are 15 rows, in fact, each with a sum of 38! Check them.

The only possible magic hexagon.
This unique figure was discovered independently by Martin Kühl of Germany about 1940 and Clifford W. Adams of Philadelphia, in 1962.

ACTIVITY

Sketch the outline of the magic hexagon in your book then insert numbers according to this condition: replace each number with the difference between that number and 20.

What is the sum of each 3 cell row? 4 cell row? 5 cell row?

Talisman Squares

They are defined as an $n \times m$ array of integers from 1 to n^2 such that the difference between any integer and its neighbour is greater than some given constant. (A neighbouring square is one that is immediately adjacent to the given square either vertically, horizontally or diagonally.)

1	5	3	7
9	11	13	15
2	6	4	8
10	12	14	16

How many neighbours has 7?
How many has 2?
How many has 11?
This talisman square has a constant greater than 1.

Would it be possible to construct a 3×3 square containing the integers 1 to 9 such that the difference between adjacent squares is always > 1?

This order 5 square is credited to Kravitz.
What is the constant by which any square differs from its neighbour?

15	1	12	4	9
20	7	22	18	24
16	2	13	5	10
21	8	23	19	25
17	3	14	6	11

Talisman Hexagons such as the one shown here, can also be devised. (Compare this with the magic hexagon.)

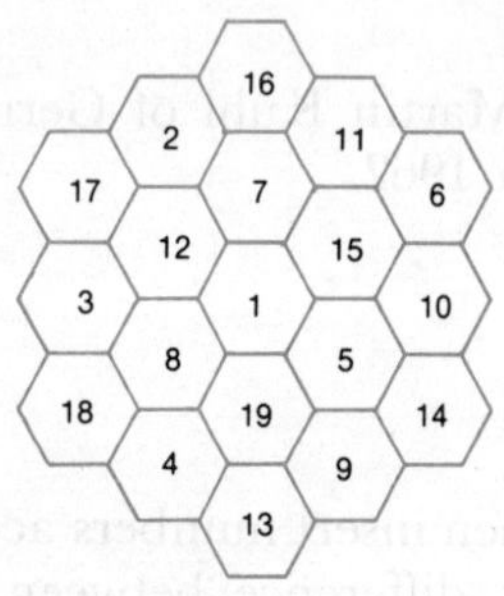

The study of talisman squares, triangles and hexagons is so new that no rules for construction nor any practical applications have been exploited fully.

A Semi-Magic Cube

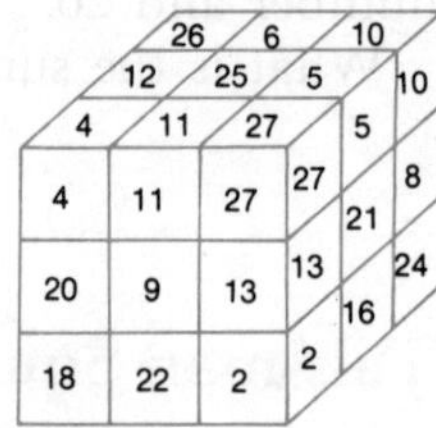

How many small cubes are in this figure? What is the magic constant on any face, either horizontally or vertically?
If the cube is broken up into its three component internal planes, the total of the main diagonals on the faces, with the exception of the central plane, is not 42. Thus it is said to be semi-perfect.

4	11	27
20	9	13
18	22	2

12	25	5
7	14	21
23	3	16

26	6	10
15	19	8
1	7	24

It has been proved that no perfect magic cubes of orders 2, 3 or 4 are possible. It is unknown if orders 5 to 7 exist.

However, a perfect order 8 cube (512 cells) was discovered in 1970 by a sixteen year old student called Richard Myers, Jr. It has a constant of 2 052 and any two numbers symmetrically opposite the centre add to 513.

Is 8 the lowest order a perfect magic cube can have? This is an open question.

Incomplete Magic Variations

There are some interesting observations that can be made when we look at the number relationships that appear on a calendar for any month of the year. As an exercise, look at the following entries for August and September, 1976.

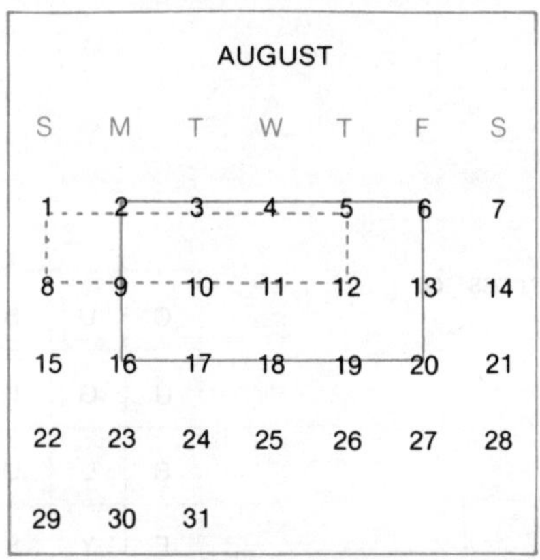

If a rectangle is drawn anywhere (excluding the blanks), the sum of the numbers at the ends of the diagonals is the same. For example, in August, 1 + 12 = 8 + 5, 2 + 20 = 16 + 6 and so on.

If a 3 × 3 square is drawn, the sum of the numbers along the main diagonals is equal (see September). Also, these sums are equal to the total of the numbers in the horizontal row and in the vertical column!

If a 4 × 4 square is drawn, the sum of the numbers along each of the main diagonals is the same.

An Interesting Game with an Incomplete 4 × 4 Magic Square

	7	2	11	18
13	20	15	24	31
5	12	7	16	23
6	13	8	17	24
10	17	12	21	28

Choose any two sets of four integers, for example {7, 2, 11, 18} and {13, 5, 6, 10}. Complete the table. Ask a friend to put a ring around any number and to cross out all the other numbers in that particular row and column. For instance, if 7 were chosen, the numbers 12, 16, 23 and 15, 8, 12 would be crossed out.

Now ask your partner to choose another number and to repeat the above rule.

Do the same a third time.

In all cases you are able to forecast the sum of the four numbers chosen, as this routine forces the sum to be the total of the numbers along either diagonal (also identical with the sum of the eight original numbers chosen to construct the square – 72 in this example).

Numerical Tick Tack Toe

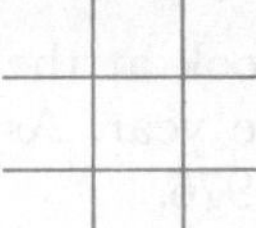

The first player uses only the odd numbers 1 to 9, inclusive, while the second player uses the evens 2 to 10 inclusive. Each in turn writes one of his or her numbers in a square. The object is to make a row, column or diagonal equal to 15.

Magic Word Squares

Here is a square with words reading across or down.

C	U	B	E
U	G	L	Y
B	L	U	E
E	Y	E	S

	E		A	
E		A		E
	A			E
A			E	
	E	E		

✱ Insert letters in the blank spaces to make a magic word square.

Unit 2 Topology-
The Mathematics of Distortion

Topology, a fairly new branch of Mathematics, is hard to define. It started as a kind of geometry but has now reached into many other fields. Generally speaking, it deals with space, surfaces, solids, regions and networks, and it is full of apparent paradoxes and impossibilities that probably make it more intriguing and provocative than any other branch of mathematics!

Topology could be called the art of analysing those permanent properties (called topological invariants) of a geometric shape that remain unaffected after the shape itself has been shrunk, stretched, twisted, crumpled or turned inside out (but not torn or broken). For instance, a doughnut will always have a hole, no matter how it is transformed.

Most topological experiments are based on **transformations**, that is, changes in the shape of a surface, without any breaking. They involve a property called a **genus**, roughly meaning the number of holes the object has.

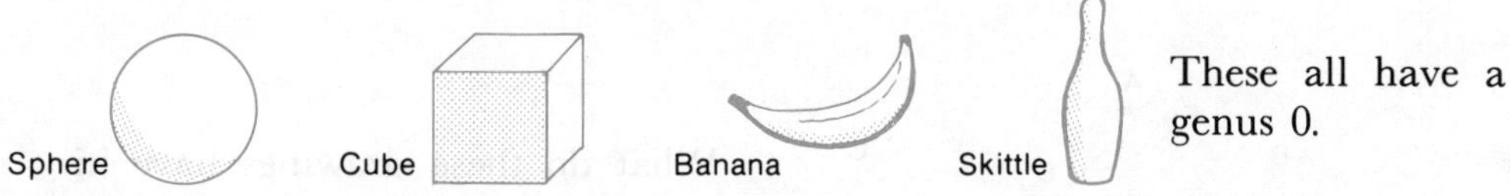

These all have a genus 0.

A doughnut (or torus) can be changed into a cup, with a genus of 1.

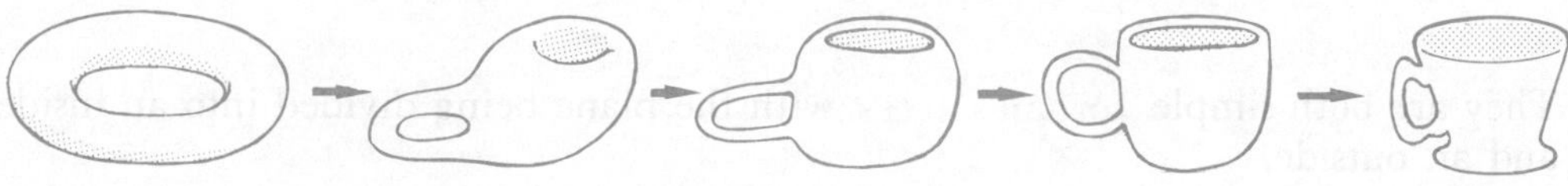

Distortions of a Cube

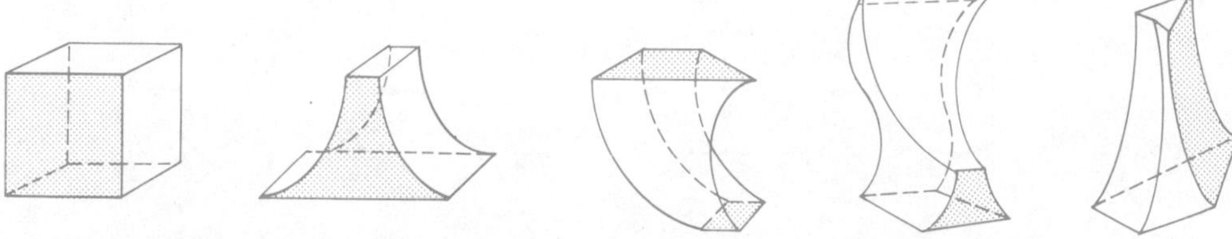

Any distortion is allowed, providing the end result is connected in the same way as the original.

EXERCISES

1. By completing the given table, classify the following objects according to their genus.

An orange, a record, a button, the figure 8, a pair of glasses, a buckle, an artist's palette, a flower pot, a jug, a piece of pipe, an egg slice, a bottle, a needle, the letter B and a telephone dial.

NAME OF OBJECT	DIAGRAM	GENUS

2. Write down some other objects which have a genus greater than 3.

Simple **closed curves** illustrate a property that is of interest in topology. (These are figures in which it is possible to start at any point and travel over every other point of the figure only once, before returning to the starting point.)

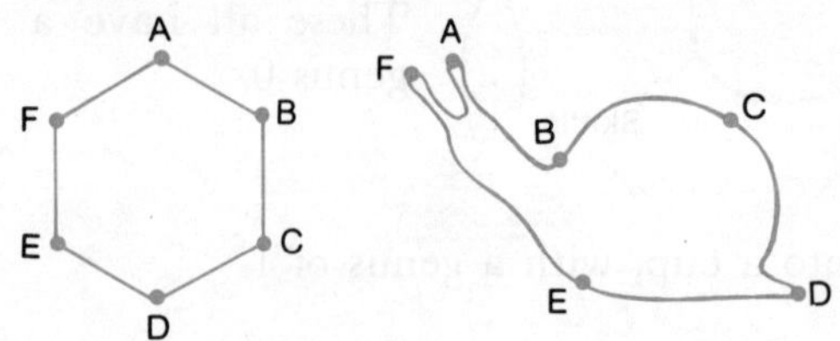

What do these drawings have in common?
What features have altered?

They are both simple **Jordan curves,** with the plane being divided into an inside and an outside.
If two figures can be twisted and stretched into the same shape without connecting or disconnecting any points, they are said to be **topologically equivalent.**
Here are some examples for you to study.

The letter I is topologically equivalent to at least ten letters of the alphabet. Can you pick them out of the following line-up?

A B C D E F G H I J K L M N
O P Q R S T U V W X Y Z

Find letters equivalent to:
1. O
2. E
3. K

Topological Invariants

An example of topological invariants comes from a theorem of major importance that was stated by the Swiss mathematician Leonhard Euler in 1752. He has been given the title of the "Grandfather of Topology", because of the immense contribution he made to the topic.

His theorem deals with polyhedra, for example, a cube and a tetrahedron (triangular pyramid) that are both solids bounded by faces (F), having straight edges (E), with the edges meeting at vertices (V).

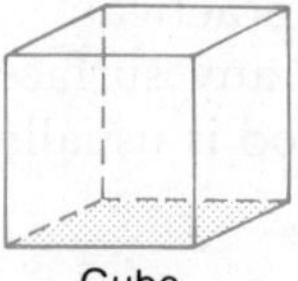

Cube Tetrahedron

How many faces, edges and vertices has each of these solids?

Euler proved that if you add the number of faces to the number of vertices, then subtract the number of edges, you always get 2, no matter how complex the polyhedron. Even a complex 240-sided solid known as a solid starred small rhombicosidodecahedron conforms with his rule. It has 240 F, 360 E and 122 V. Check these numbers in the formulae:

$$F + V - E = 2 \text{ or } F + V = E + 2$$

Rhombicosidodecahedron

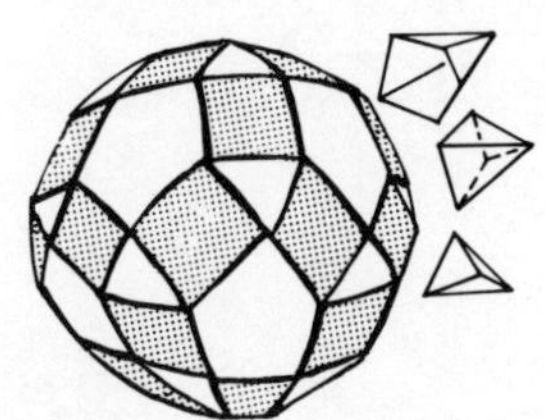

62 Faces:
20 Triangles
30 Squares
12 Pentagons

If this is stellated (with points) it becomes a 240-faced polyhedron.

Also, test the following solids:
1. a rectangular prism,
2. a square pyramid,
3. a triangular prism,
4. a frustum,

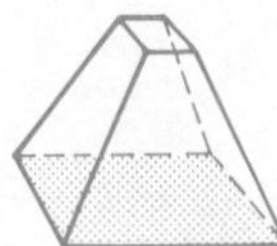

5. a truncated prism,

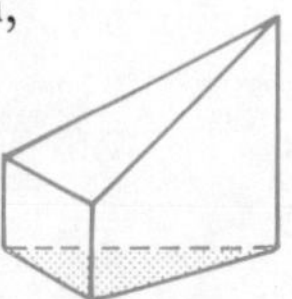

6. a truncated pyramid.

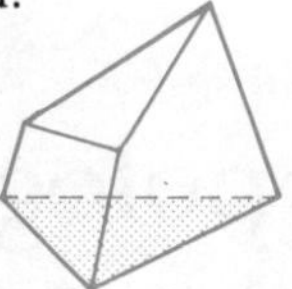

This law can be generalised to include any drawing that is in lines and dots, commencing with one dot.

$$1 \text{ F} + 1 \text{ V} - 0 \text{ E} = 2$$

Although Euler did not realise it, his curious discovery is closely linked to many network problems. Topological network theory is used today in the design of computer circuits and it has abstruse applications in astronomy. Not very practical, you might think, but then the mathematics Einstein used in his Theory of Relativity was, at the time it was created in the 19th century, rather impractical.

In topology the aim is to leave out distances entirely, so that any surface can be mapped flat, provided it is simply connected. The same method is usually followed with maps, especially those using Mercator's projection.

Networks

A network is a diagram consisting of points, called corners or junctions, joined by lines, called arcs (or routes) with the **degree** of a corner being determined by the number of arcs that finish there.

DEFINITIONS

1. A **route** is a path that can be drawn with one continuous line.
2. A route is **circular** if, when following its entire length, one returns to the starting point.
3. A **non-circular** route can either terminate (if there are two end points) or be partially circular (if there is only one end point).

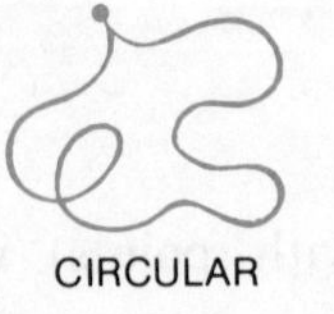

CIRCULAR

TERMINATING

PARTIALLY CIRCULAR

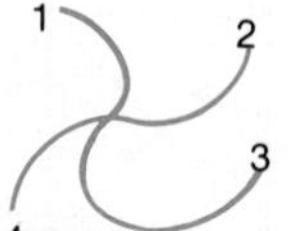

4. A junction is a point at which two or more routes meet.
5. The power of a junction is the number of routes leading from it.
6. A branch is a section of a route between two consecutive junctions.
7. A loop is a section of a route which leaves and returns from the same junction without passing through any other junctions (that is a circular section). When finding the power of this junction, the two arms of the loop count as separate branches.

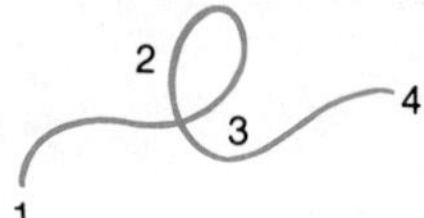

8. The order of a pair of junctions is the number of branches joining them.

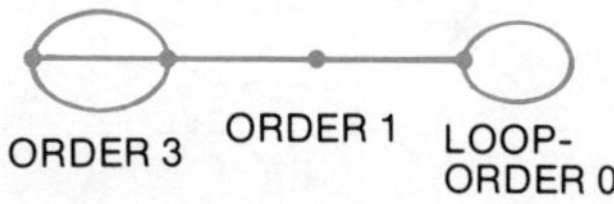

9. A network is complete if there are at least two completely distinct routes between any two junctions.

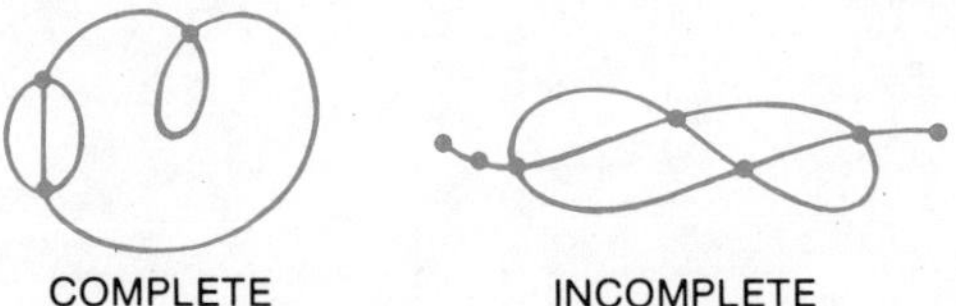

10. A region is the space bounded by one or more branches of a network.
11. The rank of a network is the minimum number of arcs needed to draw it completely if each branch is drawn just once. A rank 1 network is said to be unicursal.
12. Two networks having the same number of similarly powered junctions (occurring in the same order) are said to be equivalent.

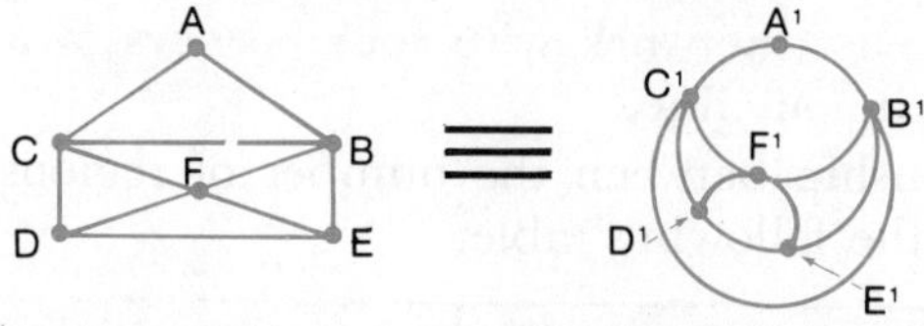

When trying to trace out the following networks in your exercise books, try to discover a rule which makes it possible to do some and not others.

You are not allowed to lift your pencil off the paper, nor retrace any arc. *Hint:* the answer to the rule depends on the degree and number of corners in the network and the place from which you start.

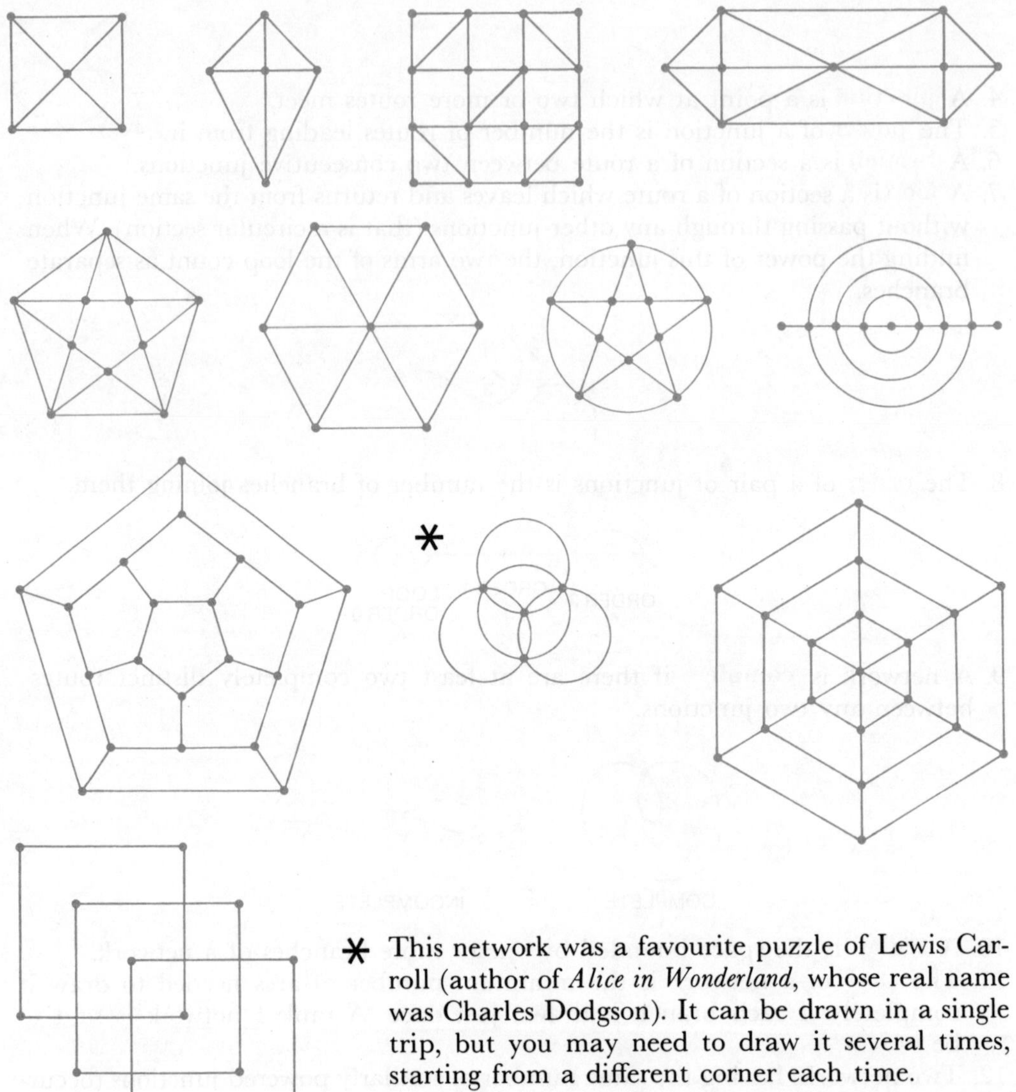

* This network was a favourite puzzle of Lewis Carroll (author of *Alice in Wonderland*, whose real name was Charles Dodgson). It can be drawn in a single trip, but you may need to draw it several times, starting from a different corner each time.

Euler was able to prove a network with four corners of an odd degree cannot be drawn in one continuous line.

To find out a relationship between the number of regions, corners and arcs in a network, complete the following table:

Network Number	No. of Regions	No. of Corners	No. of Arcs
1	3	2	3
2 to 12			

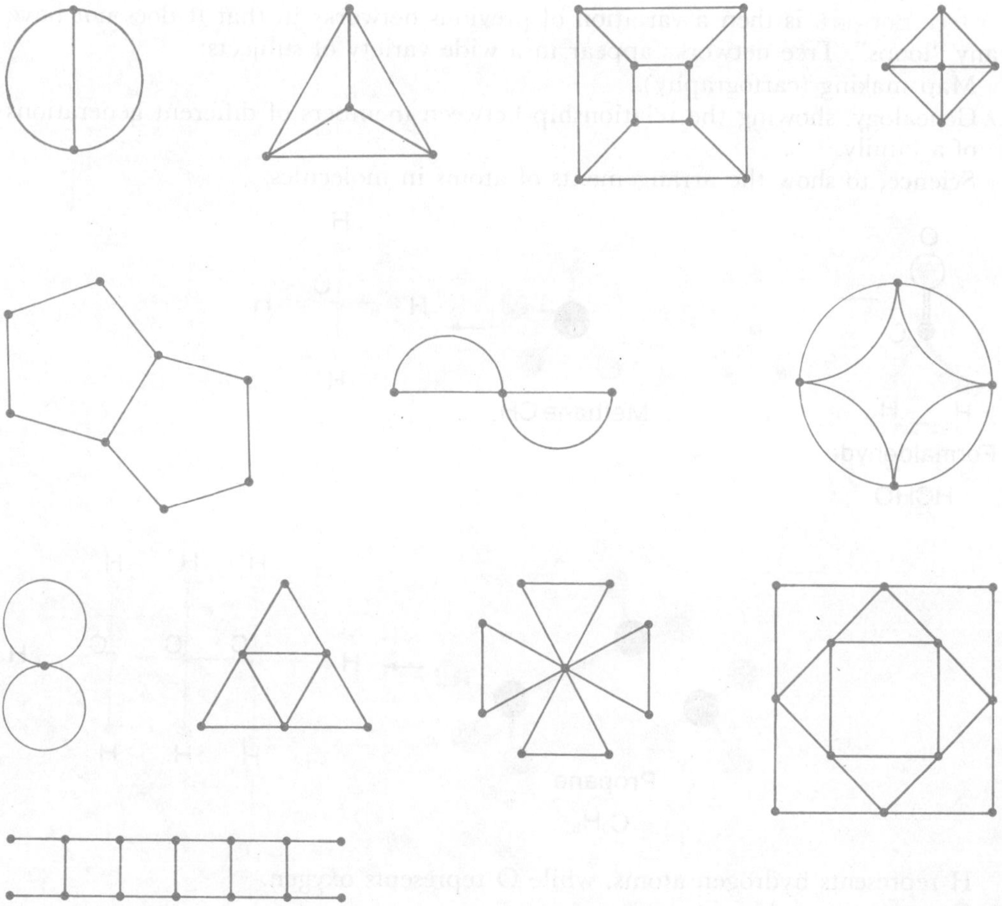

Can you write a formula connecting the number of regions (R), the number of corners (C) and the number of arcs (A)?

Does this remind you of something else that has already been mentioned?

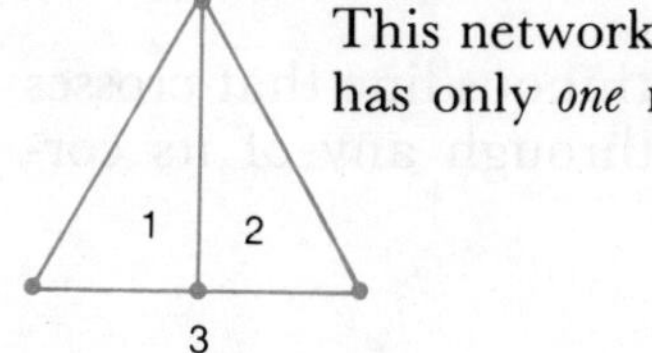

This network has three regions. Is it possible to draw a network that has only *one* region? (The one outside?)

Study this network.

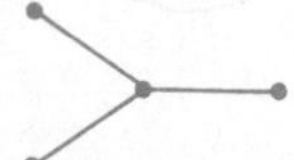

It has only an outside.

Would it be possible to trace out this network in one continuous line? The answer is, of course, no, as it has insufficient arcs or "loops".

A tree network is then a variation of previous networks in that it does not have any "loops". Tree networks appear in a wide variety of subjects:

- Map making (cartography).
- Genealogy, showing the relationship between members of different generations of a family.
- Science, to show the arrangements of atoms in molecules.

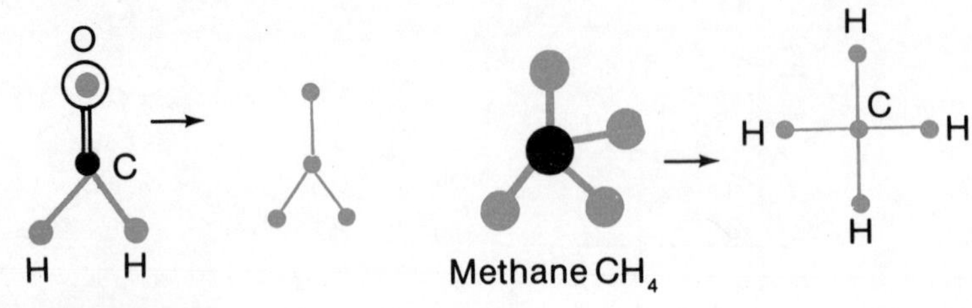

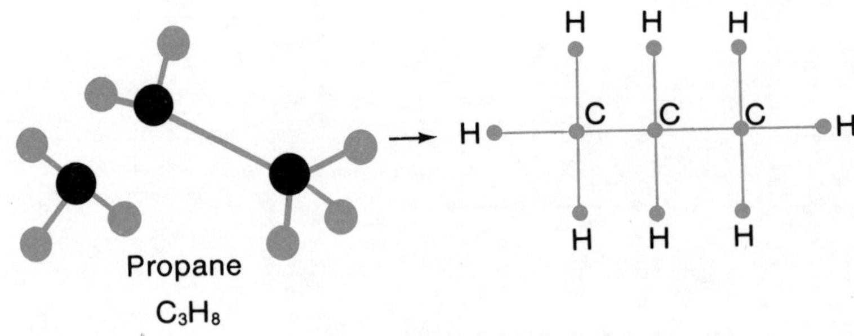

H represents hydrogen atoms, while O represents oxygen.
C represents carbon atoms.

- Probability, one of the important topics in mathematics today. Probability deals with the measure of chance, in such things as biology, horseracing and many card games. (See Unit 3 on games.)

ACTIVITY

An interesting puzzle involves the drawing of a continuous line that crosses each arc of a network only once, without going through any of its corners.

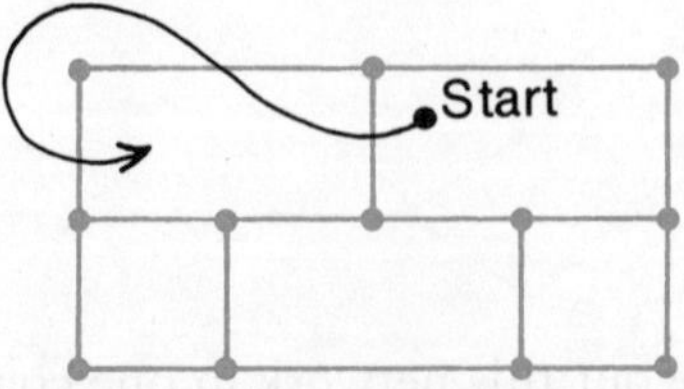

Can the puzzle be solved, or is it impossible?

Copy the following networks into your book and try to cross each arc *once* only. You may start anywhere.

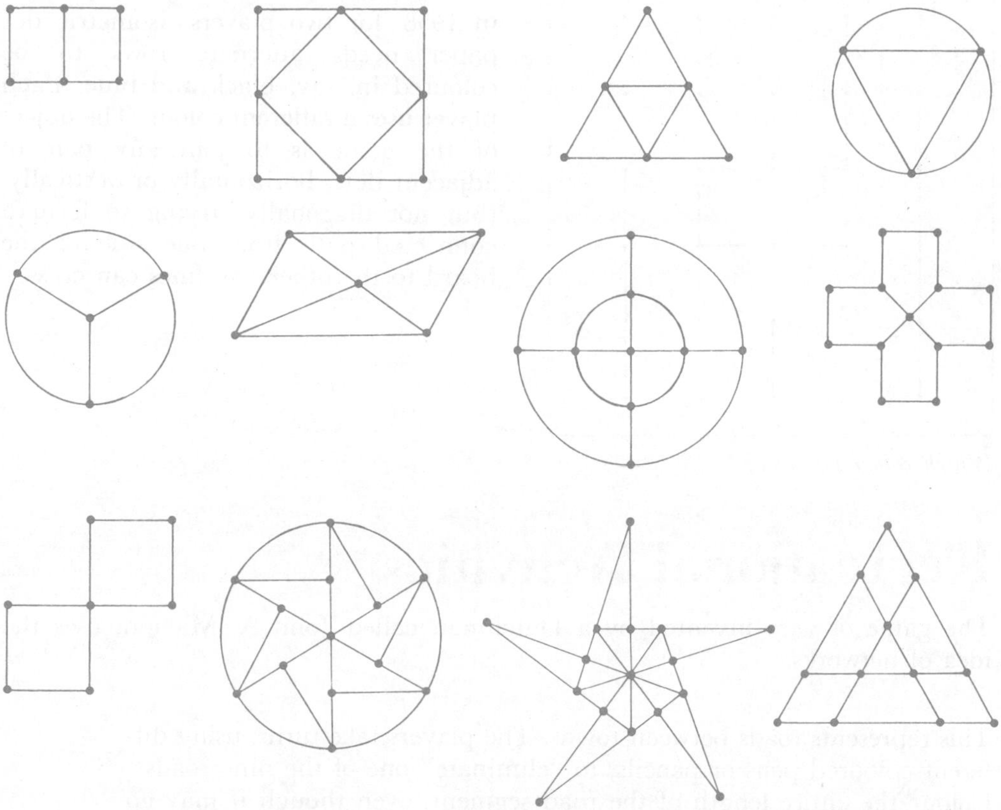

Many snowflakes have the shape of trees. From this magnified sketch draw a simple network; then see if all the arcs can be crossed.

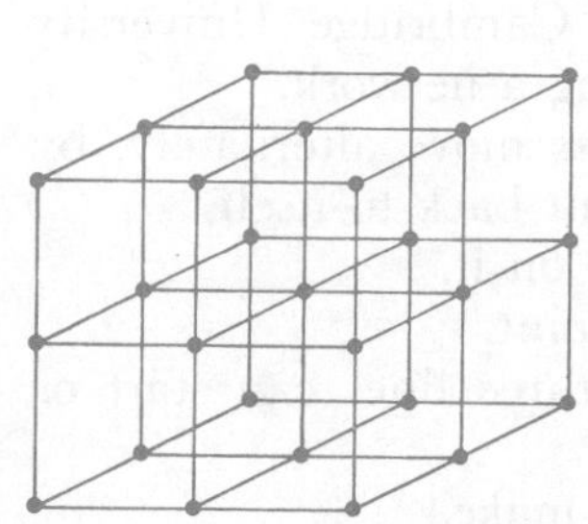

How many arcs are in this 2 × 2 cube?

A Topology Game Called Bridg-it

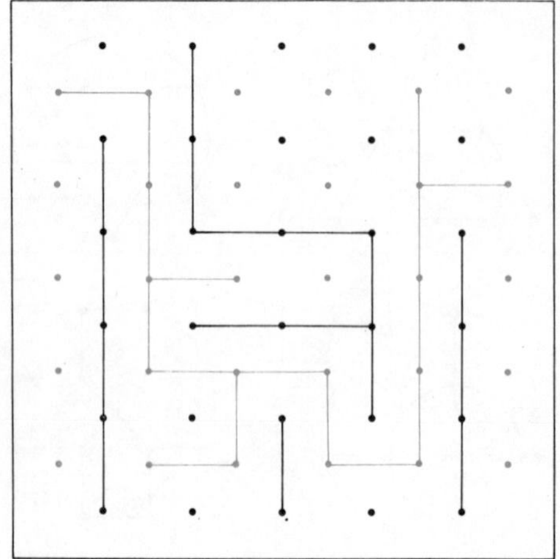

Which colour has won?

This game was devised by David Gale in 1958, for two players. Isometric dot paper needs alternate rows to be coloured in, say, black and blue. Each player uses a different colour. The object of the game is to join any pair of adjacent dots, horizontally or vertically, (but not diagonally) trying to form a connected path from one side of the board to the other. No lines can cross.

Recreational Activities

The game of Jam invented by a Dutchman called John A. Michon, uses the idea of networks.

This represents roads between towns. The players take turns, using different coloured pens or pencils, to "eliminate" one of the nine roads. Colour the entire length of the road segment, even though it may go through one or two towns.

The first to colour three roads that go to the same town is the winner.

A recent game called Sprouts was invented by two Cambridge University mathematicians. It is for two people and involves drawing a network.

Two points are marked on a sheet of paper. The players move alternately, by drawing an arc from one point to another or from a point back to itself.

When an arc is drawn, a new point is drawn somewhere on it.

No arc may cross itself, another arc, or pass through a point.

A point has only three "lives", that is, no more than three lines can start or finish at it.

The object is to leave your opponent without a move to make.

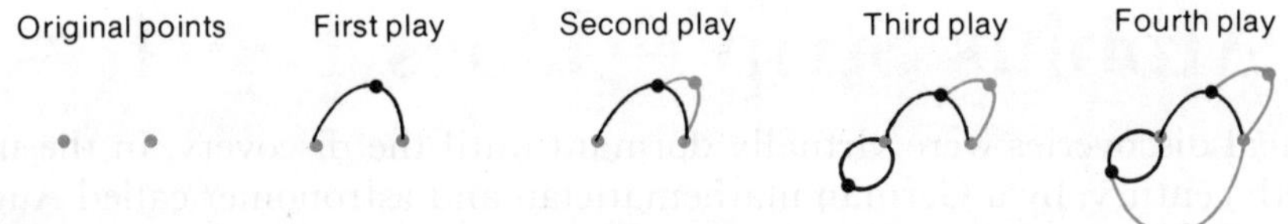

Here are diagrams showing one series of moves in an unfinished game starting with two points.

Is there a limit to the number of moves that a sprouts game can last? What conclusions can you make?

Brussels Sprouts is a variation that uses three or more points to start.

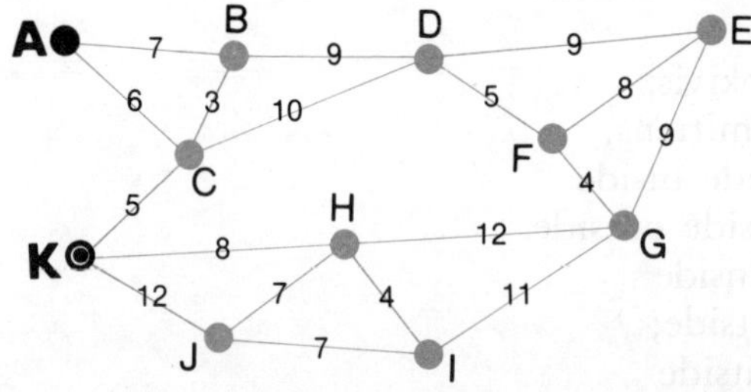

✱ What is the minimum total you can get when going from A to K, visiting every vertex once only?

✱ To which room do these instructions lead?
Enter the building. Enter the room, first door on the left. Go into the adjoining room through the door on the right. Leave this room by the door on the right. Turn left. Enter room through first door on the right. Go through door straight ahead.

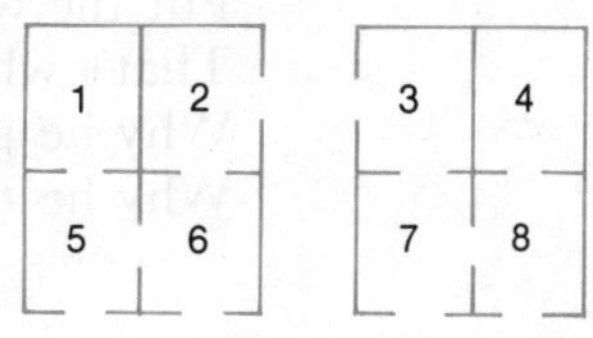

9 10 11
12 13 14

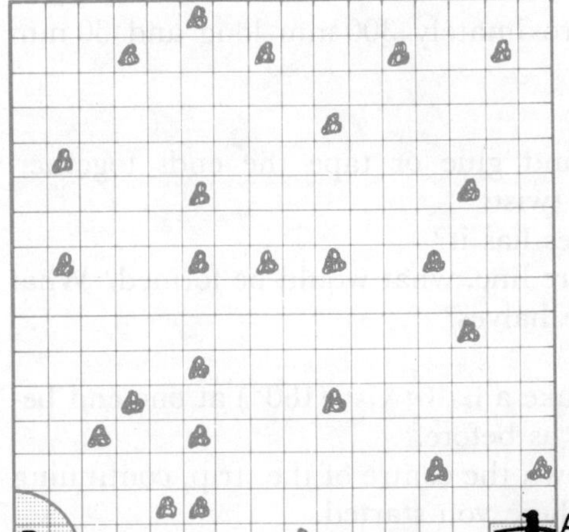

These symbols represent dangerous rocks. A boat must not venture within one square, in any direction, of the rocks or it will be swept to its doom.

How can the fisherman navigate his boat from the fishing grounds A to port B safely?

The Möbius Strip — Does 1 + 1 = 1?

Topological discoveries were virtually dormant until the discovery, in the middle of the 19th century, by a German mathematician and astronomer called Augustus Ferdinand Möbius (1790–1868) of a one-sided strip of paper.

This limerick summarises some of the striking properties of the Möbius Strip.

A mathematician confided
That a Möbius band is one-sided;
But you'll get quite a laugh,
If you cut one in half,
For it stays in one piece when divided.

Consider this: He killed the noble Mudjokivis,
Of the skin he made him mittens,
Made them with the fur side inside
Made them with the skin side outside.
He, to get the warm side inside
Put the inside skin side outside;
He, to get the cold side outside
Put the warm side fur side inside
That's why he put the fur side inside,
Why he put the skin side outside
Why he turned them inside outside.

From a book of *Humorous Verse,*
compiled by Carolyn Wells.

To understand what happens when a Möbius strip is cut, it is necessary to carry out the practical experiments yourself and to record the results in a table. It is hard to predict the outcomes!
Cut about eight strips of paper, each approximately 300 mm long and 30 mm wide.

Step 1. Take one of the strips and glue or tape the ends together, making sure there are no twists.
How many sides and edges has it?
If it were cut down a centre line, what would be formed? What are the properties of these halves?

Step 2. Take another strip and make a half-twist (180°) at one end before joining the two ends, as before.
Now draw a pencil line down the centre of the strip, continuing until you come back to where you started.
Cut along this line. What happens?
How many half-twists are there in the model now?

Step 3. Take another Möbius strip and cut along it parallel to an edge and about one-third of the way from the edge.
Continue cutting until you arrive at your starting point.
What is the result this time?
What are the dimensions (length and breadth) of these loops?

Step 4. Do the above operations again, except that you must cut it one-quarter of the way from an edge.
In what ways is this result similar to or different from the previous one?
Can you guess what the result would be if you cut around the strip one-fifth of the way from an edge?

Repeat steps 2 to 4 on strips with two half-twists (360°) three half-twists (540°), and four half-twists (720°).
Summarise your results.

NUMBER OF HALF TWISTS	CENTRE CUT FORMS.....	DESCRIPTION IN WORDS	SKETCH	ONE-THIRD CUT FORMS	ONE-QUARTER CUT FORMS
0	Two separate strips	Half as wide, same length as original			
1 (one surface, one edge)	One strip	Half as wide, twice as long			
2 (two sides, two edges)					
3					
4					

What do you think the result would be if you took a strip with twenty half-twists in it and cut it down the centre?

Rules may be formulated from these experiments.
If n is even, two strips similar to the original result, linked in the same way as the boundary curves.
If n is odd, only one strip results, similar to the boundary curve. That is, for $n \geq 3$, it is knotted. It has $2n + 2$ half-twists and two extra, gained when the coils are opened out. (n for each circuit of the original.)
If the strips are trisected, the centre strip will resemble the original, but the outer ones will be single or a pair (like the result of bisection) and they will be linked to the centre ring.

Variations on the Theme . . .

1. Cut a strip of paper about 250 mm long by 50 mm wide, then cut two slits at each end (as shown in the diagram).

A	X
B	Y
C	Z

Form a loop by joining A to X, pass B over A and glue to Y, which is passed under X.

Similarly, pass C between ends A and B then pass Z over X and join.
What happens when the cuts are continued all the way around the band?

2. Prepare another strip in exactly the same way, labelled the same, but this time join A to X, B, over A to Y, then C over both B and A, to Z. Cut, as before.
What is each of the bands now? Cut one and see what happens.

3. Prepare a third piece of paper in exactly the same way. Turn C over (give it a half-twist) and join it to X. Turn A over and join it to Y. Turn B over and join it to Z.
Complete both cuts. Did you expect this result?

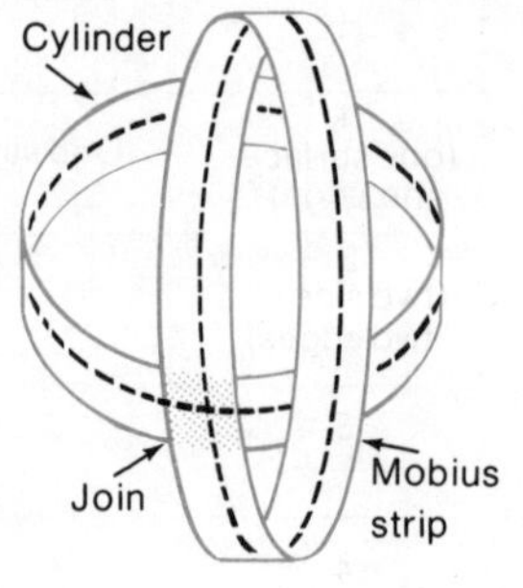

Pseudo-Möbius Strip

Cut two paper strips, approximately 240 mm by 30 mm.
Join the first to form an open cylinder.
Now make a Möbius strip with the second.
Attach securely, as shown.
Remember: The cylinder has two edges and surfaces while the Möbius strip has only one edge and one surface.

Cut along the dotted lines.

Now what happens? What shape does it become?
This final form has two sides, two edges and no twists! (The twists cancel themselves in the unfolding.)

Slit Strips

With a long piece of paper, cut a slit before pushing one end through and joining. Extend the slit right around the band (as in diagram).
In a similar manner, push the top end through the slit after giving it a twist, towards you, so that when the ends are joined, the top surface is glued to the bottom surface.
What happens?

Siamese Möbius Strips

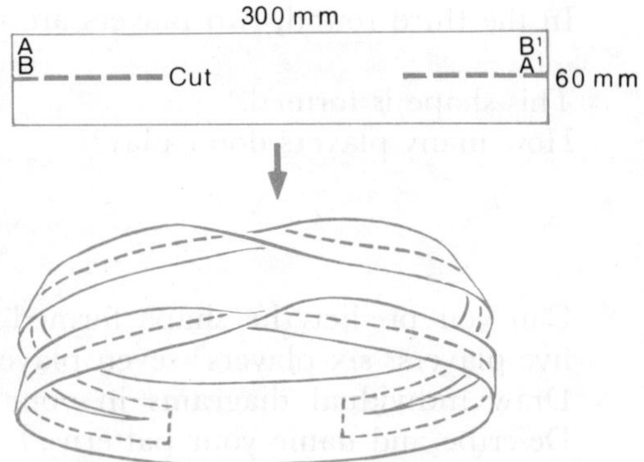

Take a long strip of paper and cut two longitudinal slits, as in the diagram. Bring the upper pair of ends together and join with a half-twist so that A joins A', and B joins B'.
Do the same with the lower slits, but twist in the opposite direction. Then cut along the dotted line.

RESEARCH

In the book entitled *The Graphic Work of M. C. Escher* (Pan/Ballantine, 1973, plate 40) there is a Möbius network strip with nine red ants crawling after each other, endlessly. It is called "Möbius Strip II". See if you can find it.

Only recently physicists found a use for the Möbius strip. By devising electrical circuits in this form they discovered that they could get rid of unwanted electrical inductive "resistance" at low temperatures, so making an electrical perpetual motion machine a practical reality.

Electrical circuits are such that specific points on components must be joined by wires, but the exact lengths and shapes of the wire connections are not determined by the needs of the electrician, so that the geometrical arrangement of a set of components may be modified without disturbing the electrical circuit. In fact, some internal connections in radio and TV sets are made with flexible wire so that parts may be moved after the wiring has been completed. This is essentially a topological arrangement since it does not involve sizes, lengths, straight lines or ratios which enter metrical and projective geometries.

The R. F. Goodrich Company has also patented a long-lasting rubber conveyor belt in the form of a Möbius strip!

Basketball Networks

Suppose twelve players are arranged in a circle. In the first round, the ball is thrown in a clockwise direction, to every player. What is the locus of the ball?

In the second round a player is skipped, that is, the ball goes from 1 to 3 to 5 to 7 to 9 to 11, then back to 1.

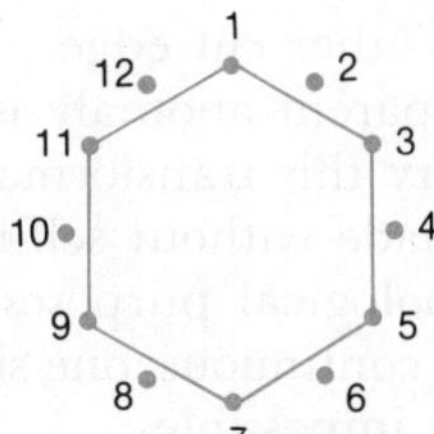

What is the name of this figure?
How many players miss out?

In the third round, two players are missed.

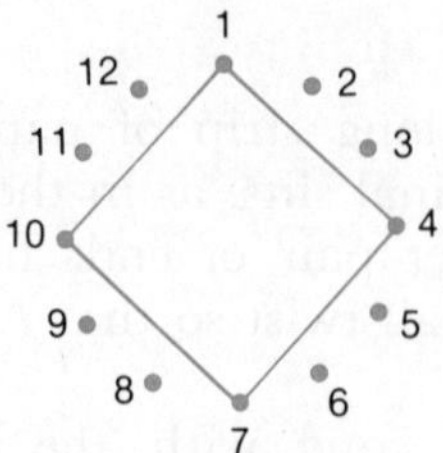

This shape is formed.
How many players don't play?

Can you predict the shape formed if three players are skipped? four players? five players? six players? seven players?
Draw individual diagrams in your book and trace out the patterns formed. Describe and name your patterns.
In how many do all players participate?

✳ Belts

These wheels are connected by belts as shown.

- If A rotates clockwise, can all four wheels also rotate?
- If so, which way does each one turn?
- Will the wheels turn if all four belts are crossed or if one or three are crossed?

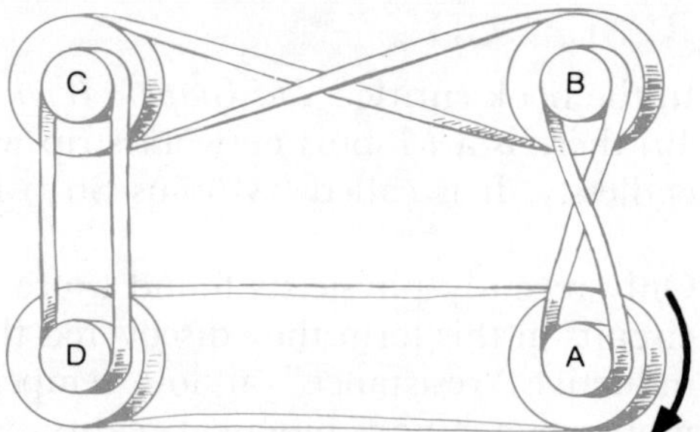

The Klein Bottle

Another one-sided phenomenon is the Klein bottle, invented in 1882 by the great German mathematician Felix Klein (1849–1925). It is one-sided, no-edged, has an outside, but no inside.

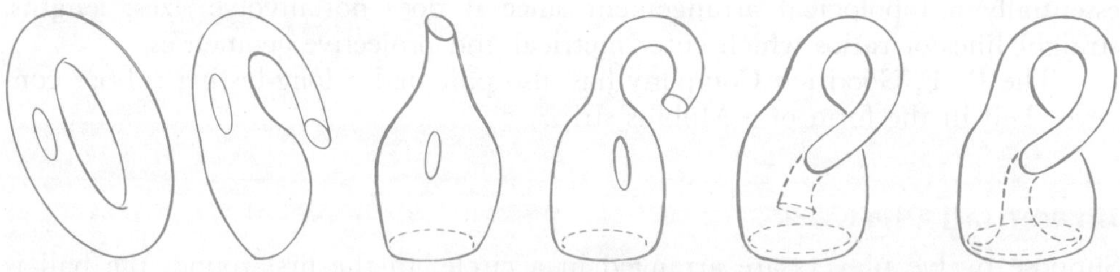

Imagine a torus being cut once longitudinally, then the resulting "tube" being stretched, distorted and flared so that one end passes through the bottle and joins up with the other cut edge.

The apparent anomaly is that the resulting bottle has no edges!

In theory this transformation is possible, but in practice, no such bottle has ever been made without self-intersection.

For topological purposes one usually supposes that no hole actually exists so that the continuous one-sided surface passes through itself. This, of course, is physically impossible.

What happens if a Klein bottle is cut into two longitudinal sections? Does it, like a Möbius strip, remain in one piece but lose its one-sidedness?

Diagrammatically, it yields two odd-shaped pieces, both Möbius surfaces, one right-handed and one left-handed (or mirror images).

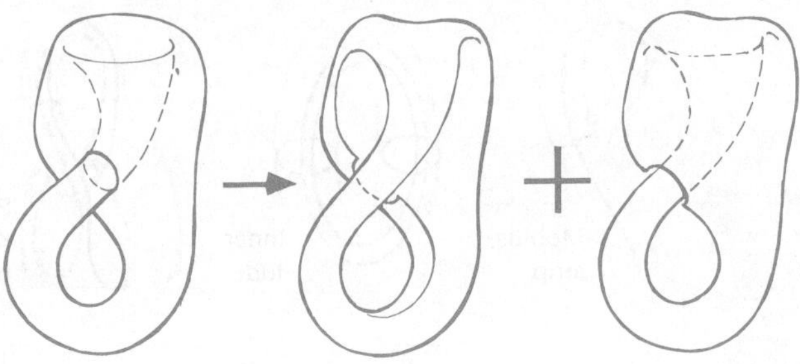

A Klein bottle may be simplified and transformed onto a sheet of paper.

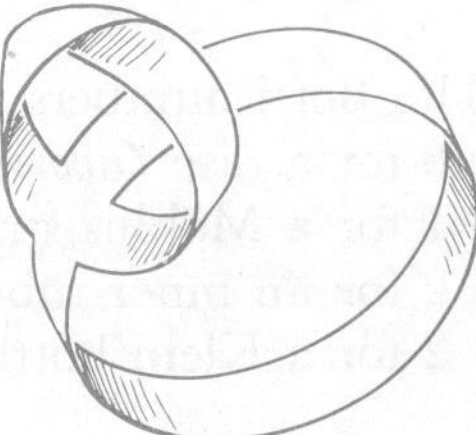

Each topological type of surface has several distinguishing features of a surface that characterise it, namely:

- the number of edges;
- the number of sides;
- its Betti number;
- its chromatic number.

Betti Numbers

These are named after the 19th century Italian physicist Enrico Betti and are the maximum number of cross cuts that can be made on a surface without dividing it into more than one piece. A cross cut may be thought of as a simple cut with scissors; it begins and ends on the edge. Any cross cut on a disc will divide it into two pieces, so that its Betti number is 0.

The Betti number of a Möbius strip is 1, while a Klein bottle or inner tube has a Betti number of 2.

In the case of solids, the number is always one less than the number of faces. A tetrahedron, therefore, has a Betti number of 3 (or 4F − 1). What would be the Betti number of a hexagonal prism?

Loop cuts provide another way of finding the Betti number of a surface. These cuts begin and end at a point in the surface, avoiding the edge entirely. By counting the maximum number of loop cuts that can be made without dividing a one-edged surface into more than one piece we again get the Betti number of the surface. Cross cuts and loop cuts were used by the German mathematician Bernhard Riemann in 1857 to define the connectivity of a surface. He called a disc "simply connected", a band or halo "doubly connected" and so on, his connectivity number always being one more than the Betti number.

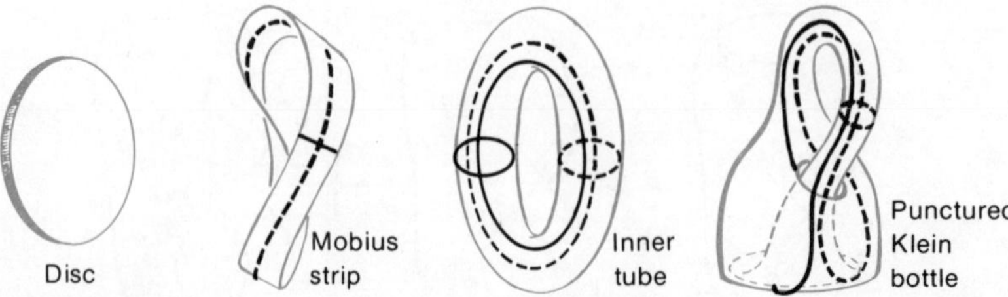

- Cross cuts (solid lines) begin and end at the edge.
- Loop cuts (dotted) avoid edges.

The Betti numbers are:
- 0 for a disc *(two sides, one edge);*
- 1 for a Möbius strip *(one side, one edge);*
- 2 for an inner tube *(two sides, one edge);*
- 2 for a Klein bottle *(one side, one edge).*

The Chromatic Number

This is the maximum number of regions that can be drawn on the surface so that each region has a border in common with every other region.

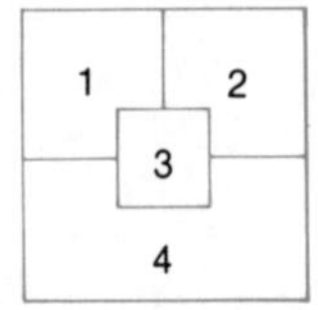

A square's chromatic number is 4.
A sphere has the same number.

Also, the term indicates the minimum number of colours necessary to colour a finite map on a plane surface. In the case of the Möbius strip and Klein bottle, the chromatic number is 6 while the torus has one of 7!

String and Rope

Innumerable tricks and stunts with string, exploited by at least three generations of magicians, may be regarded as topological. Most employ a piece of cord with knotted ends that forms a simple closed curve. This loop can be twisted around the fingers in certain ways until it seems hopelessly entangled, then a pull forces it easily away.

These loops, as early as 1904, were called "Afghan bands". See what you can find out about them.

Knots

This is a subject also taken up by topologists but not much has been proved beyond saying that a knot cannot exist in more than three dimensions. They are all basically the same – just loops or circles. However, it has been established that knots may be either right-handed or left-handed!

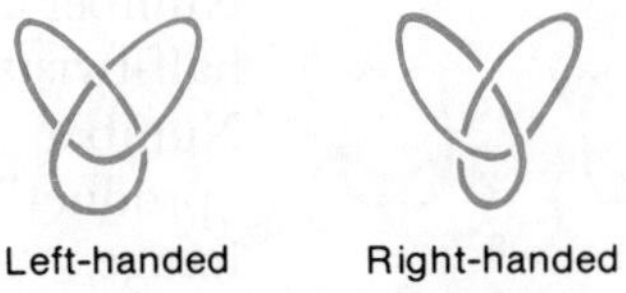

Look carefully at the following diagrams:

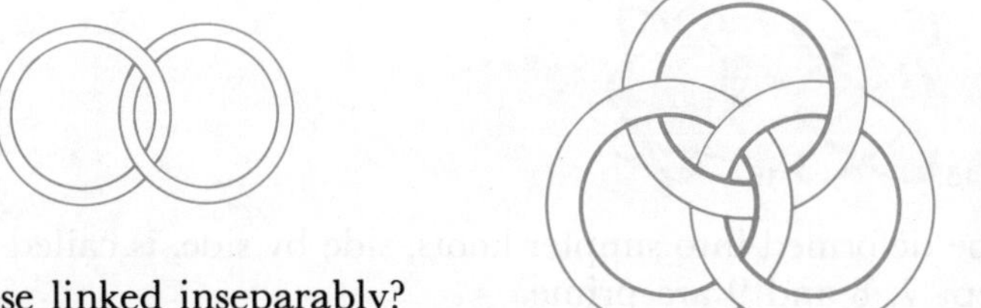

Are these linked inseparably?

None of these is linked, yet they are inseparable!
What would happen if one were cut?

They are called **Borromean rings** since they form the arms of an Italian family called Borromeo.

They are also the trade mark of an American liquor manufacturer!

This is a **Trefoil Knot** that may be compared with a Möbius strip with three half-twists, if the strip is flattened and a tracing made around the path of its edge.

Escher used knots in some of his works. His fascination with topology is expressed in such works as the woodcut "Knots" (1965) where there are two mirror-image forms of the trefoil knot.

Escher's wood engraving "Three Spheres" first appears to be a sphere undergoing progressive topological squashing. However it is not.

Find a picture of these works and investigate their properties.

Today, the study and classification of knots involves advanced analysis and mathematics of the highest order.

Here are illustrations of knots with four crossings (1), five crossings (2 and 3), six crossings (4 to 8), seven crossings (9 to 16).

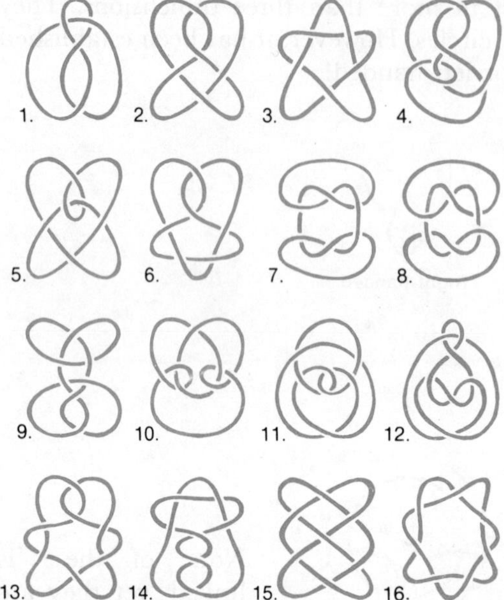

Number 3: Give a strip of paper five half-twists and join the ends.
Number 7: This square knot is the "product" of a trefoil and its mirror image.
Number 8: "Granny" is the product of two trefoils of the same handedness.
Number 16: How many half-twists to make this model?

A knot that cannot be deformed into simpler knots, side by side, is called a prime knot. All these, except 7, 8 and 9 are prime.

One might suppose that living organisms would be free of knots, but this is not so. Thomas Brock, a microbiologist at Indiana University reported in *Science,* Vol. 144 no. 1620 (15 May 1964), on his discovery of a string-like microbe that reproduces by tying itself in a knot which pulls tighter and tighter until the knot fuses into a bulb and the free ends of the filament break off to form new microbes.
Do humans ever tie part of their anatomy in knots?

MAP COLOURING PROBLEMS

Some topology "theorems" have been propounded and proved (for example, Jordan curves), but one remains notorious and has done so for over one hundred years. It is known as the **colour map problem** and states that only **four colours** are necessary to colour any map, provided it is drawn on a plane or even a curved surface. The theory of map-colouring is reputed to have been begun by Möbius in 1840 as one of the problems he set for his students.

This is a simple diagram illustrating the necessity for four colours.

A number of regions can meet at a **point** but are not considered to be touching. For instance four regions on a chess board meet at a single point, yet only two colours are used, not four.

No one has made a map that needs more than four colours and no one has been able to prove that such a map can be made!

In how many essentially different ways can a cube be coloured with only three colours?

When we delve into some of the topological curiosities such as the Möbius strip and the torus (doughnut), however, a different set of "rules" applies.

It has been shown that a Möbius strip needs six colours while a torus requires a maximum of seven.

Draw an enlargement of this in your exercise book then try colouring it, using the minimum number of colours.

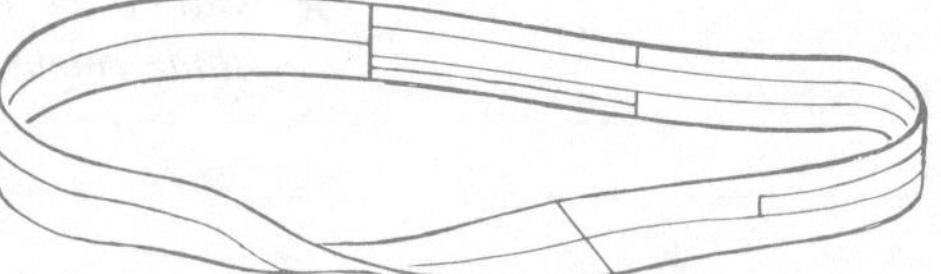

Imagine a torus flattened out into a rectangle. The "map" shown is in this form; it consists of seven polygons, each of which touches the other six.

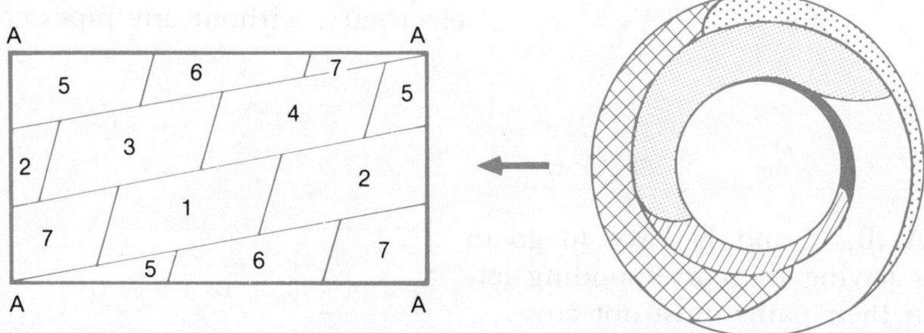

If it is practicable to turn a torus on a lathe, the map itself can be marked out and coloured. The "waist" of the torus is divided into three, by three points. A helical curve is then drawn on the torus, which begins at one point, travels all around the torus and returns to the next point and so on, until after three circuits the curve returns to its starting point. Divide the whole length of this curve into fourteen equal parts. Then join the first point of subdivision to the sixth, the third to the eighth, the fifth to the tenth and so on until every even point, $2n$, is joined to the odd point $2n - 5$.

The result will be seven elongated areas, each in contact with the remaining six.

Teasers

What is the minimum number of colours needed for these "maps"?

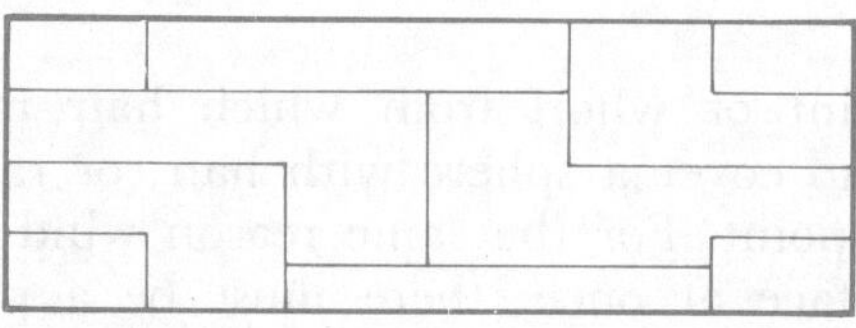

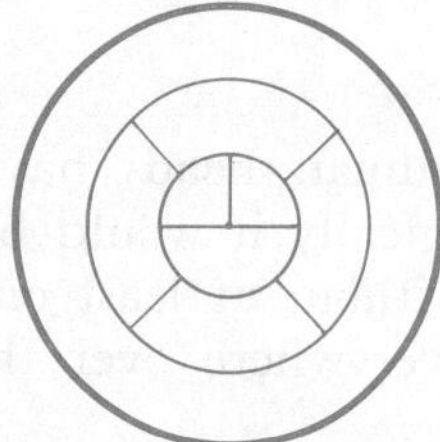

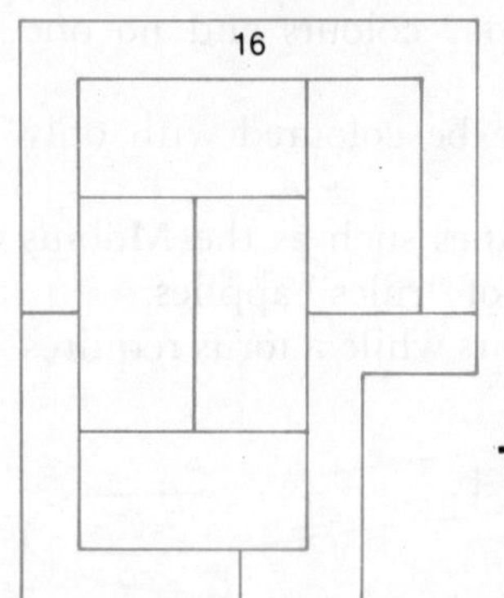

All the regions in the diagram have an area of $8m^2$ except the top one, which has an area of $16m^2$.

There are only enough colours to paint –

$8m^2$ blue

$24m^2$ red

$24m^2$ yellow

$16m^2$ green

✱ *Can it be done so that no two adjacent areas are the same colour?*

Topological Tomfoolery

 1.

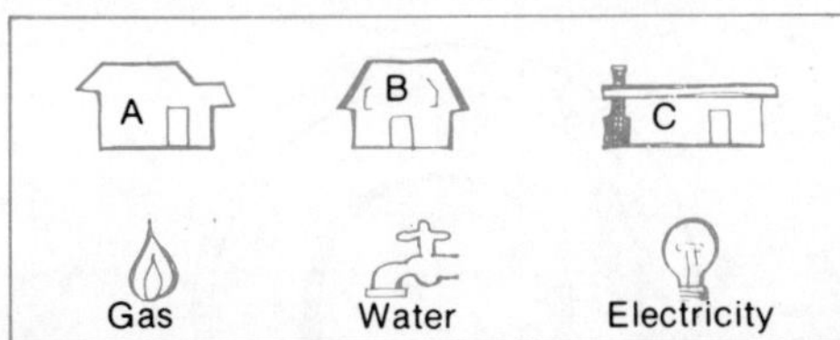

Connect each house with gas, water and electricity, without any pipes crossing.

2. Boys A, B, C and D have to go to schools having the corresponding letter, but their paths must not cross.

   Can you do it?

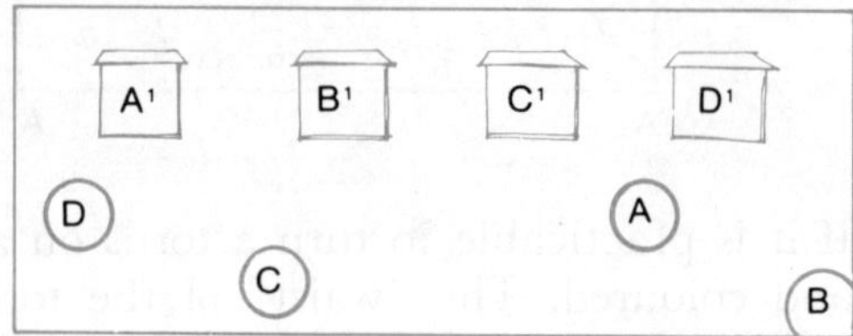

✱ 3. How can you tie a knot in a piece of string *without letting the ends go?*

✱ 4. A party trick.

   Tie a loop of string to the end of a fairly long pencil making sure the pencil is longer than the loop.

   How can you attach the pencil to a button hole on a coat or cardigan?

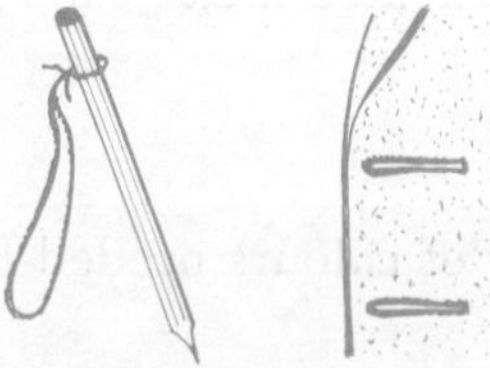

As a matter of fact . . .

Most human heads have a fixed point or whorl from which hair radiates. Topologically it would be impossible to cover a sphere with hair, or radiating lines, without at least one such fixed point. For the same reason wind cannot blow everywhere over the earth's surface at once; there must be a point of calm.

Schlegel Diagrams

1. Cut the top face off.

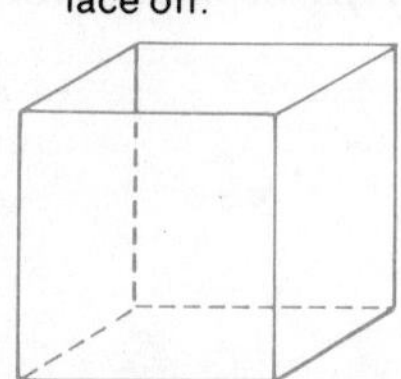

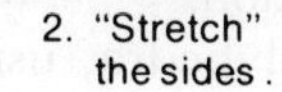

2. "Stretch" the sides . . .

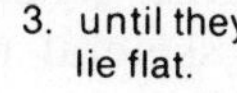

3. until they lie flat.

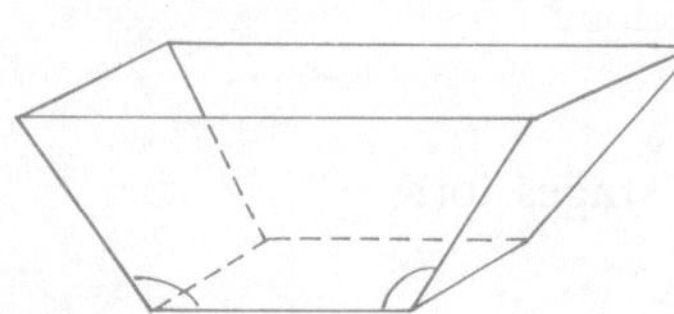

All the faces, edges and vertices can be seen at once.

The Five Platonic Solids

PERSPECTIVE VIEW | NET | SCHLEGEL DIAGRAMS

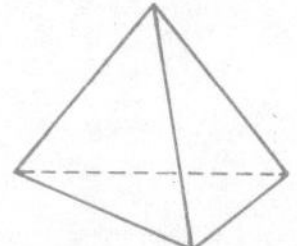

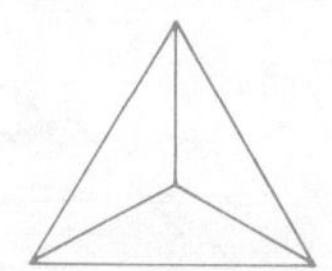

TETRAHEDRON {3, 3} This means 3 vertices and 3 shapes at each vertex.

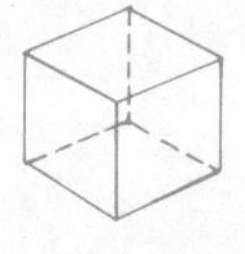

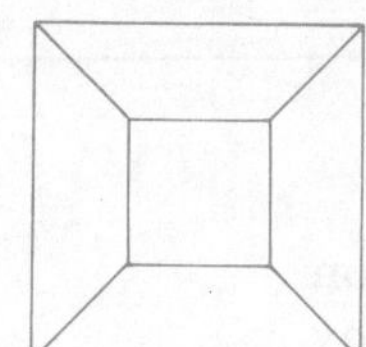

HEXAHEDRON {4,3} (cube)

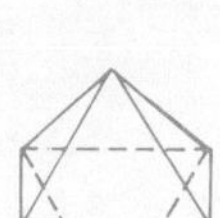

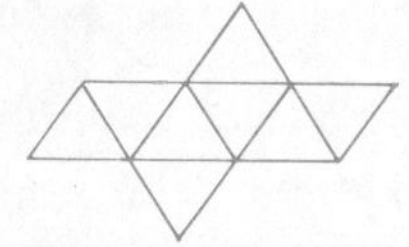

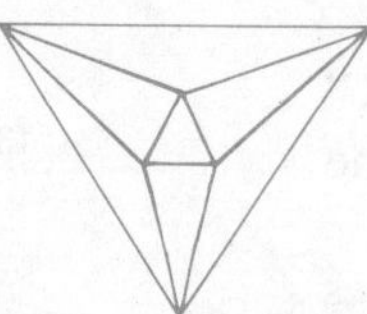

OCTAHEDRON {3,4}

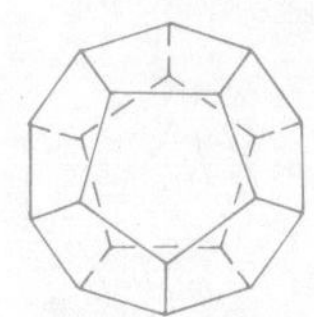

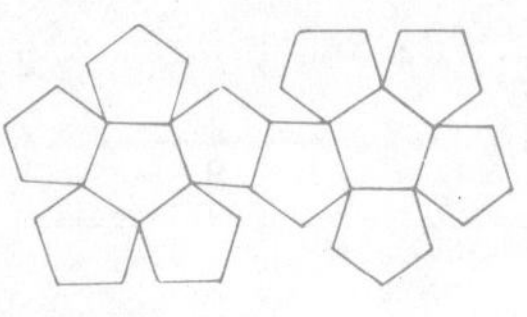

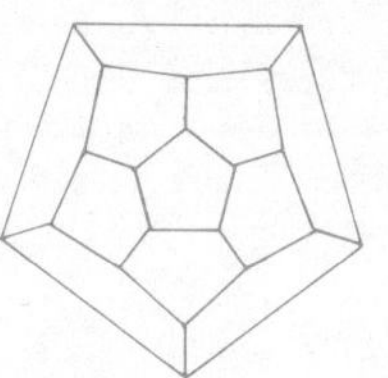

DODECAHEDRON {5,3}

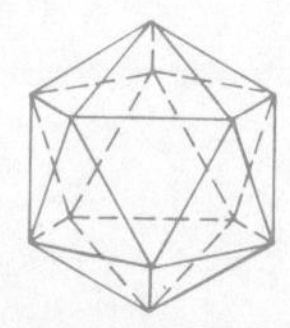

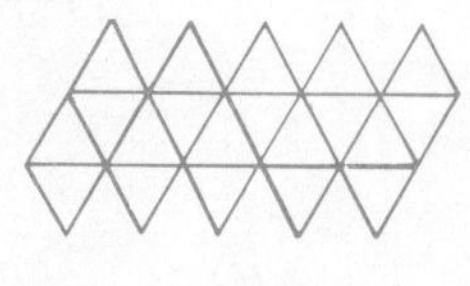

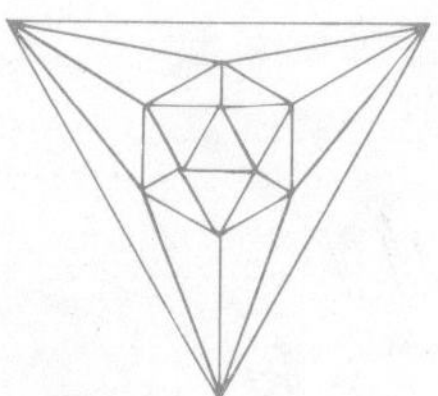

ICOSAHEDRON {3,5}

Solid	V.	E.	F.	Dihedral Angle
Tetrahedron	4	6	4	70° 32′
Hexahedron	8	12	6	90°
Octahedron	6	12	8	109° 28′
Dodecahedron	20	30	12	116° 34′
Icosahedron	12	30	20	138° 11′

A pair of icosahedral dice of the Ptolemaic dynasty can be seen in one of the Egyptian rooms of the British Museum in London.
Leonardo da Vinci made skeletal models of polyhedra, using strips of wood for their edges, and left the faces to be imagined.

ACTIVITY

1. Try drawing the above three stages for:
   - a rectangular prism
   - a triangular prism
   - a pentagonal prism
   - a square pyramid
   - a triangular pyramid (tetrahedron)
2. Using cardboard make four or five stages and glue them together (see illustration for a cube).

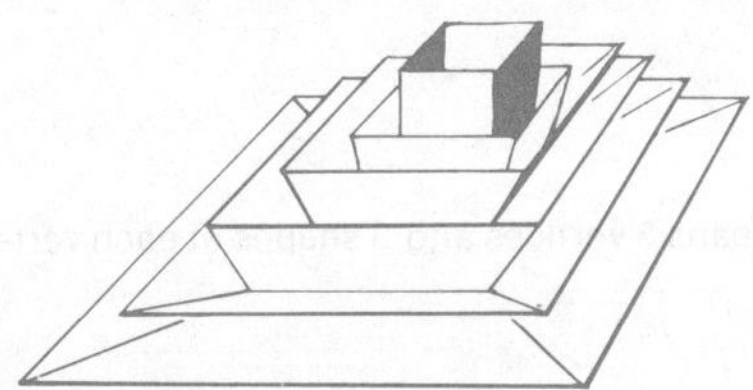

A Game to Play

Object: To place your mark on four or more sides of any hexagon, thus winning its point value.

How to play: Players mark "X" or "O" on any of the sides of the thirteen hexagons.

The winner: Player who reaches 62 points or more, first.

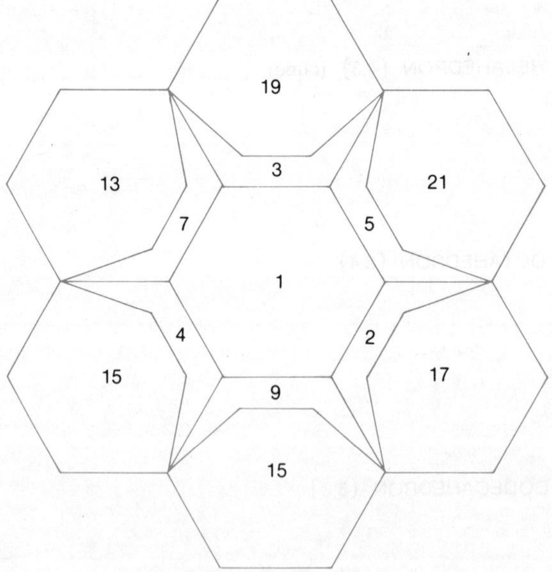

Unit 3 Pastimes

Billiards

The name probably originated from the French word *bille*, meaning a stick. Although the origin of the game is uncertain, it was known in France in the 15th Century and in England in the reign of Elizabeth I.

The introduction of cues, about 1820, giving much greater control of the ball, made possible the modern development of scientific billiards which is largely dependent on the principles of geometry and dynamics.

Two outstanding masters in the history of billiards have been the Australians Walter and Horace Lindrum.

Today, skilful players can determine where a ball will go before it is actually hit. The path, or locus, is determined by how the ball is hit, the shape of the table and the positions of the other balls. The actual dimensions of tables can vary, but they are always rectangular and twice as long as they are wide. A standard table is approximately 3.65 m by 1.82 m and is 0.86 m high.

Lewis Carroll, the author of *Alice in Wonderland*, was very interested in the mathematics of billiards, so he invented a version of the game that was played on a circular table!

Anyone who has played billiards or pool may have noticed how the cue ball can come to a complete stop after striking another ball. For this to happen the motion of the moving ball must be along the line of centres of the two balls and must not have too much rotational velocity (spin). For an illustration of this principle, try the following experiment with a number of coins. Repeat it ten times.

Paradoxes like the one illustrated in this experiment can be explained by the use of the physical concept of momentum, that is, the product of the object's mass, (m) and its velocity (v), in the formula

$$\text{momentum} = m \times v$$

The law of the conservation of momentum states that the sum of the momenta of two bodies before collision is the same as the sum of their momenta after collision. Look up a science book for further information on this topic.

Suppose a ball is hit from the corner of a table so that it travels at a 45° angle with the sides of the table. *Where will it go?*

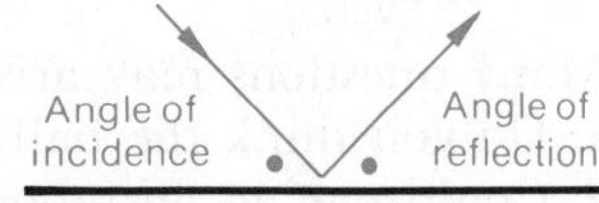

To determine the path of the ball, it is necessary to know that when the ball strikes the side of the table (cushion), the angle of reflection is equal to the angle of incidence, that is, it rebounds at the same angle as it hits.

This fact is useful to know in games like tennis and billiards. A beam of light acts in the same way: rays of light are reflected in periscopes, binoculars and other such instruments.

Here are some grid "tables" which will make it easier for you to work out the path of the ball. (There is only one ball on the table.)

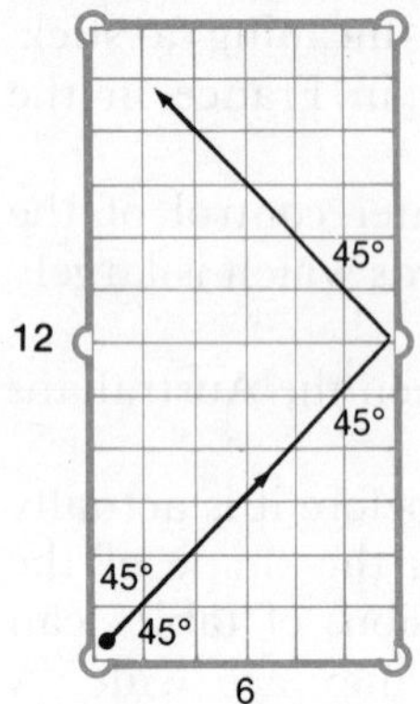

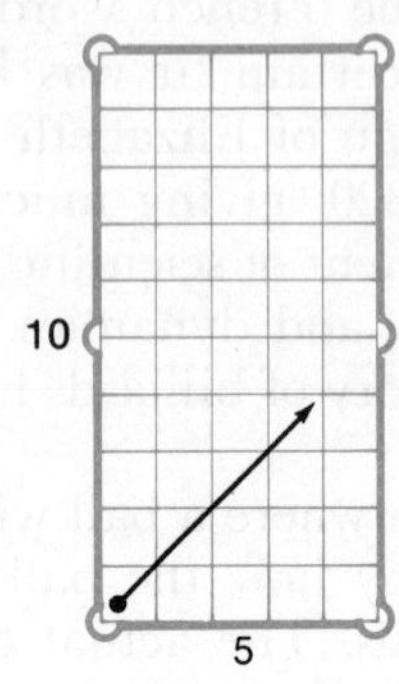

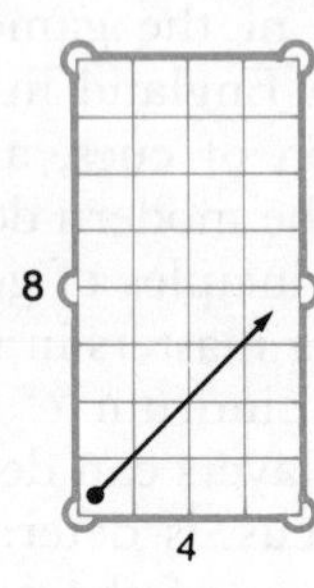

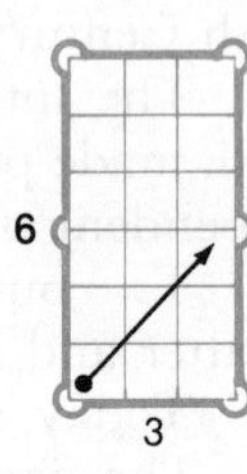

As these tables are all **similar** (the corresponding ratios of the sides are the same:

$$\frac{12}{6} = \frac{10}{5} = \frac{8}{4} = \frac{6}{3} = 2)$$

the path of the ball will be the same, ending in the top left-hand corner in each case.

Let's find out the path on a "table" with a different shape.

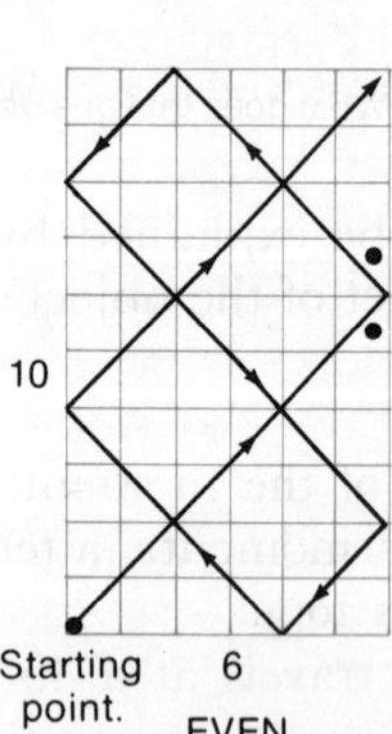

Follow the paths in each diagram. You will notice that the ball goes to a different corner, after many rebounds (called hits).

What are the simplest ratios for the length : width of these tables?

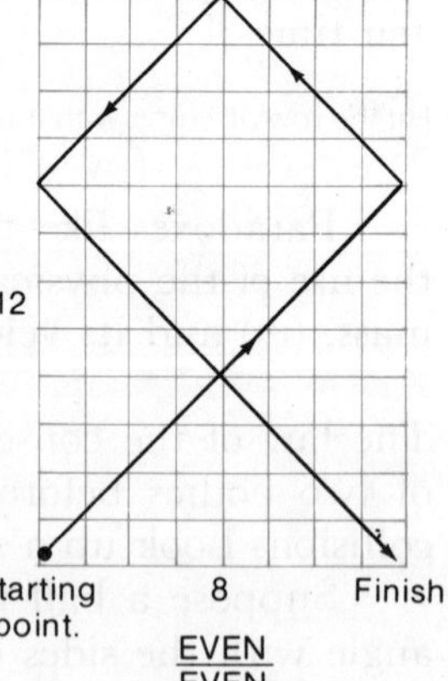

Many questions may arise.

1. Do you think the ball will always end up in one of the table's corners?
2. Could one go on rebounding from the cushions indefinitely?
3. Could one ever come back to its original corner?
4. Can you predict which of the three corners the ball will end up in if the dimensions of the table are known?

To find out the answers to these questions, do these exercises carefully on 5 mm grid paper, then paste the "tables" into your exercise book. Write the dimensions on the sides and try to formulate some simple rules as you go along. The ball always starts in the bottom left-hand corner and stops when it reaches a corner.

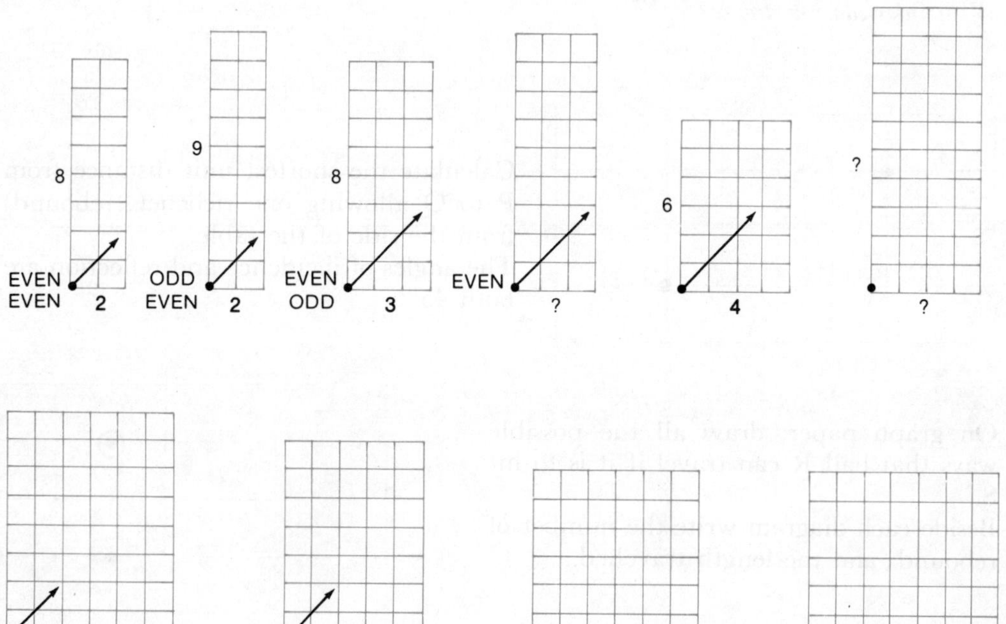

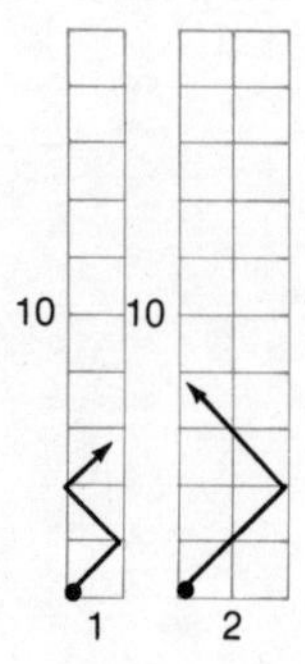

11. Draw a set of eight billiard "tables" with lengths of 10 units and widths that vary from 1 to 8 units. Put a red cross at the place where the ball ends.
   - Are there any surprises?
   - Which table has the simplest path?
   - What are the dimensions of the table with the most complex path?
   - Where does the ball finish on the tables with a width that is an odd number?

12. This time draw a set with lengths of 7 units and widths from 1 to 7. Carry out the same operations.

Can you think of a rule for telling in advance on which tables the ball will travel over every square? *Hint:* Think of factors of numbers.

From all the exercises write down the ratios of the lengths to the widths of the tables where **the ball ends up in the lower right-hand corner.** Reduce these ratios to their simplest forms.
Repeat these steps for all the tables where **the ball ends up in the top right-hand corner.**
Do the same for those ending in the **top left-hand corner.**
What rules can you deduce?

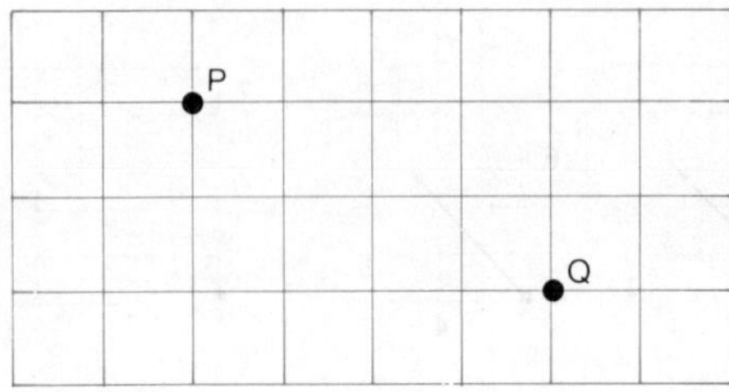

Calculate the shortest unit distance from P to Q allowing one ricochet (rebound) from the side of the table.
The angles of incidence and reflection are both 45°.

On graph paper, draw all the possible ways that ball R can travel if it is to hit S.
Beside each diagram write the number of rebounds and the length travelled.

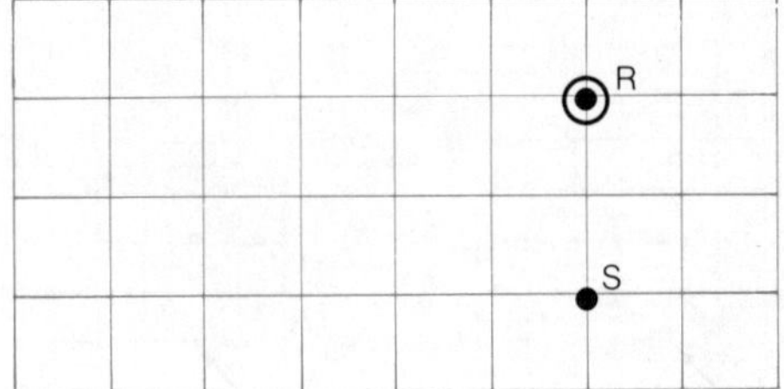

A pool table has six pockets: one at each corner and two at the midpoints of the longer sides.

In this diagram, will the ball go into the pocket marked A?

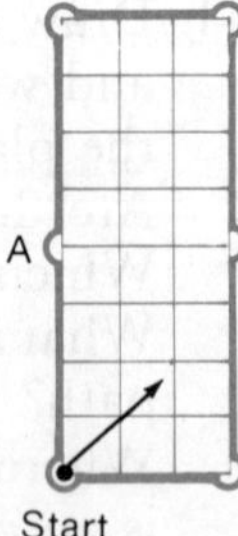

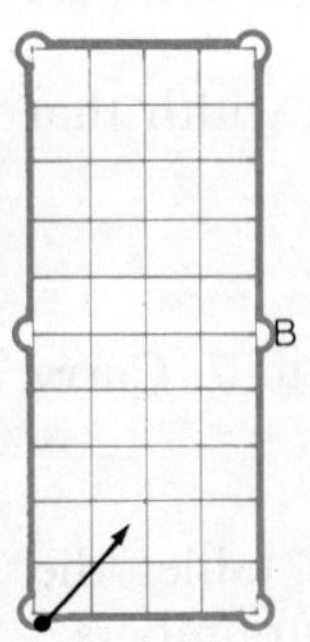

Will this ball go into pocket B?

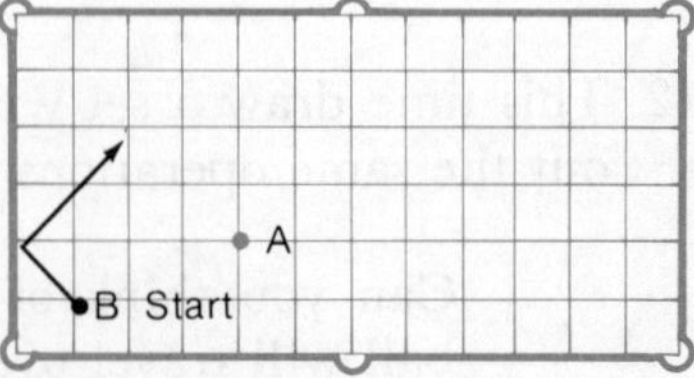

Will ball B hit ball A?

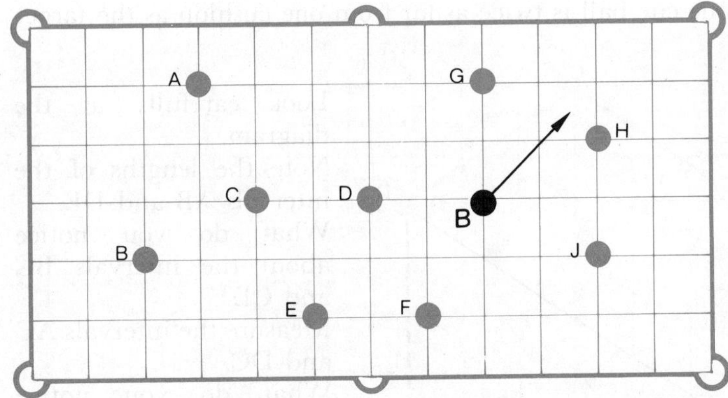

Which of the labelled balls will B hit?

✱ Where must the ball be placed on the long side of this table so that when hit at a 45° angle, it falls into a corner pocket after four rebounds?

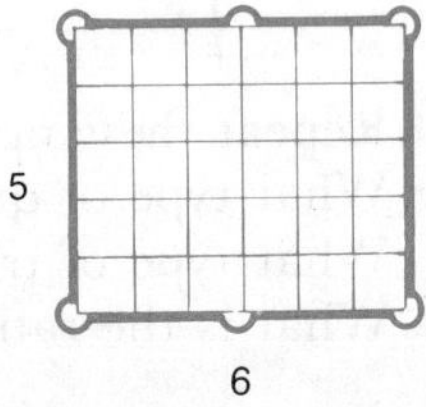

In all the diagrams so far, the ball has struck and rebounded off the cushion at an angle of 45°. What happens if it approaches at a different angle?

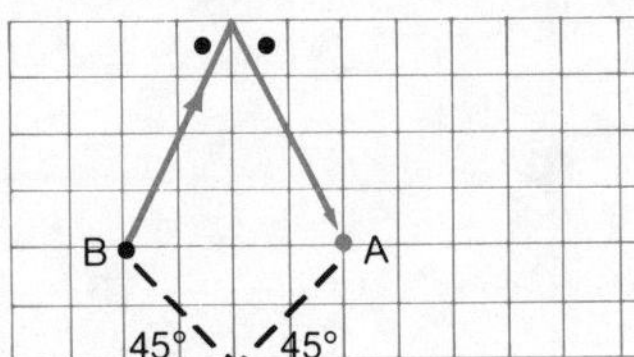

Here is a diagram which shows two ways in which the cue ball, B, can hit ball A (*if no direct hit is possible*). What do you notice about the points where the ball strikes the cushions? The dotted line makes a 45° angle with the cushion. What angle does the other path make?

• Do you know any way in which this angle may be found?

EXERCISES

1. What do you notice about the points of contact with the cushions?
   Use a protractor to measure the angles marked P, Q, R and S.

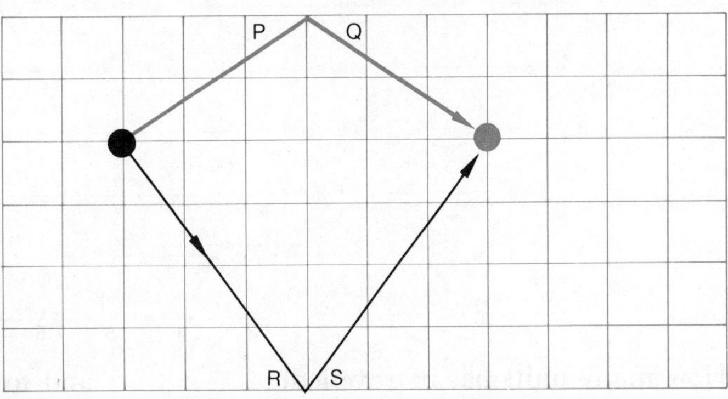

2. What happens when one cue ball is twice as far from one cushion as the target ball?

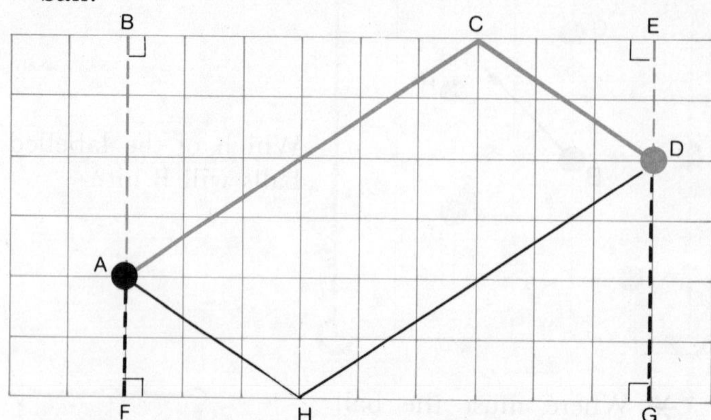

Look carefully at the diagram.
Note the lengths of the intervals AB and DE.
What do you notice about the intervals BC and CE?
Measure the intervals AC and DC.
What do you notice about each pair?

- Repeat these operations with AF and DG, FH and GH, and AH and DH.
- What type of quadrilateral is ACDH?
- What type of triangles are ABC and DEC (or AFH and DGH)?
- What is the ratio of the area of the triangle ABC to the area of DEC?

A Teaser

Trace a path taken by the cue ball, so that half the remaining balls are inside, and half outside, its path.

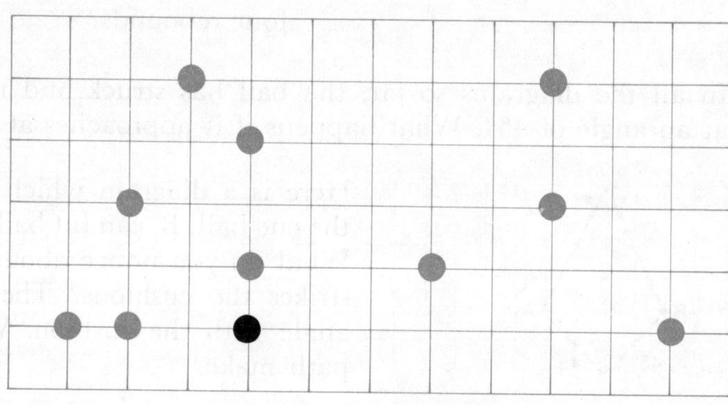

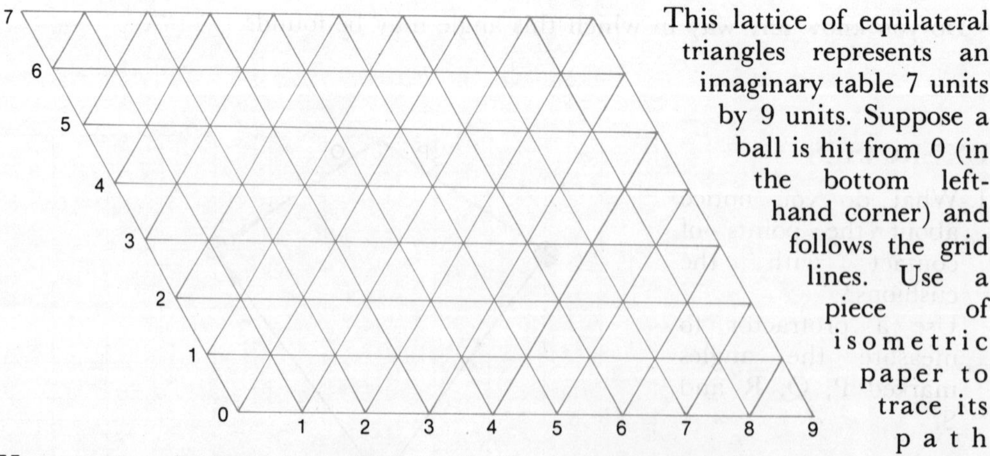

This lattice of equilateral triangles represents an imaginary table 7 units by 9 units. Suppose a ball is hit from 0 (in the bottom left-hand corner) and follows the grid lines. Use a piece of isometric paper to trace its path and to find out where it will finish.

How many units has it travelled?

Elliptical pool tables have been made. They went on sale in the USA in 1964 under the name of **Elliptipools.** There is only one pocket which is at a point called the **focus.**

Because of the elliptical shape of the table, a ball that is hit in any direction from the focus, opposite the pocket, will end up in the pocket!
This diagram illustrates the fact.

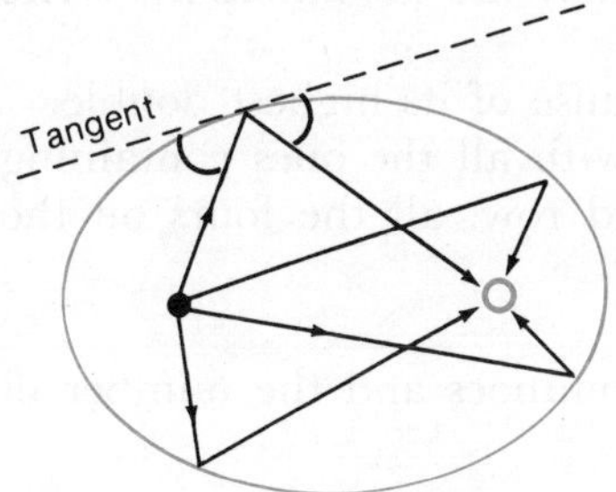

As before, the ball rebounds from the cushion at the same angle it hits it. (We have to imagine a tangent drawn at the point of contact in order to get equal angles.)

If the ball is hit between the foci, it travels endlessly along paths that never get closer to the foci than a hyperbola with the same foci.

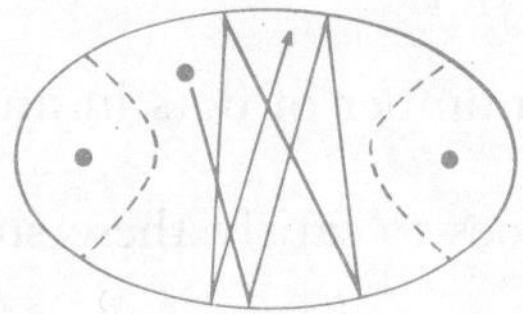

Suppose the table were **round** and the cue ball was hit from the centre of the table. What would be its locus? What relationship has this path with the tangent drawn to the circle at the point where the ball hits the cushion?

Further Games

"Man has never shown more ingenuity than in his games." *Leibniz.*

A Flutter on the Horses

To win the "double" means to pick the first place getters in the two feature races of the day. When betting on the "quinella", a punter correctly chooses the first two place-getters in the race.

Assuming there are fifteen horses in a race, the chance of picking the first place-getter is 1 in 15 (or odds of 14 to 1 against). The probability of picking the "double" is 1 in 15^2 or 1 in 225 (odds of 224 to 1 against).

The quinella has better odds. If again there are fifteen horses in the race, and the winner is correct, the chance of picking the second place-getter is 1 in 14, so you have a probability $\frac{2}{15} \times \frac{1}{14}$. This means the quinella's odds are 2 in 210 (15×14) or that the odds are 104 to 1 against.

Dice

How many different number combinations are possible on four dice, so that the total is 14?

Dominoes

A full set of dominoes has all combinations from a double blank to a double six. How many dominoes are there in a full set, if they are all different? Write them down in an orderly arrangement.

The normal set we use is called a *double six* because of its highest double.

If all the dominoes are arranged in a triangle with all the ones containing a six on the bottom row, all the fives on the second row, all the fours on the next row and so on, how many are there altogether?

- How many dots are in this set?
- What is the connection between the number of dominoes and the number of dots?

Now remove all the dominoes that have a six.

- How many are left in the *double five* set? How many dots?

  By continuing to remove a set at a time, work out the connection between the number of tiles and the number of dots. Record it in a table.
- Do you recognise a number pattern found in Pascal's triangle? (See next unit.)
- How can you calculate the number of dots in any set if you know the number of dominoes?
- Arrange the relevant dominoes to satisfy these simple arithmetical problems.

Example: 500 would be

```
 500
   2
1000
```

Try arranging these:

```
  6      554      33
  6        6     536
 12     3324     653
                1222
```

- What do these represent?

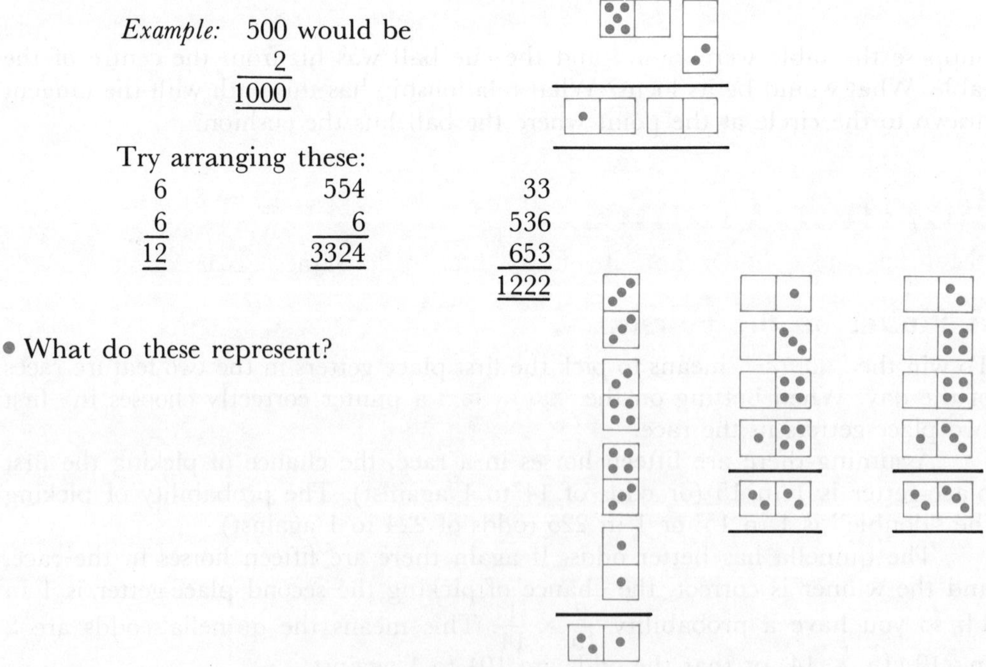

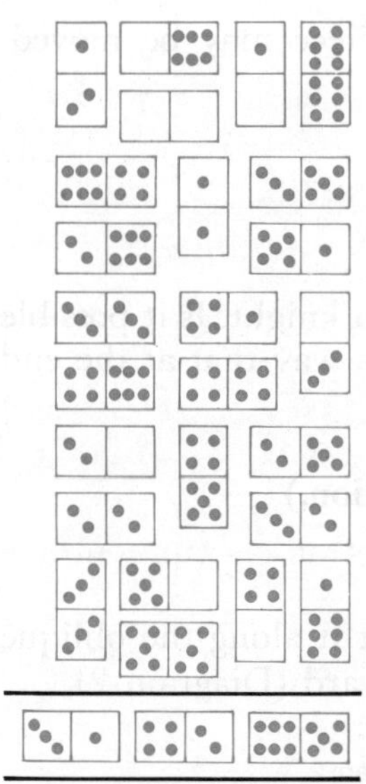

The full set is used in this mammoth addition. Convert it into numbers, add it and see if you agree.

Use a full set of dominoes to make some symmetrical arrangements. Use 1 cm grid paper. Some patterns could have holes in them.

Remove all the dominoes with blanks and doubles (fifteen should remain). These can be regarded as fractions.

+ + + + = $2\frac{1}{2}$

Arrange five of them to form a sum of 10 (improper fractions such as $^4/_3$, $^6/_1$ can be used.)

- If all the dominoes are placed in a continuous chain (adjacent ends of the dominoes must match) so that four dots are at one end, how many dots will be at the other end?

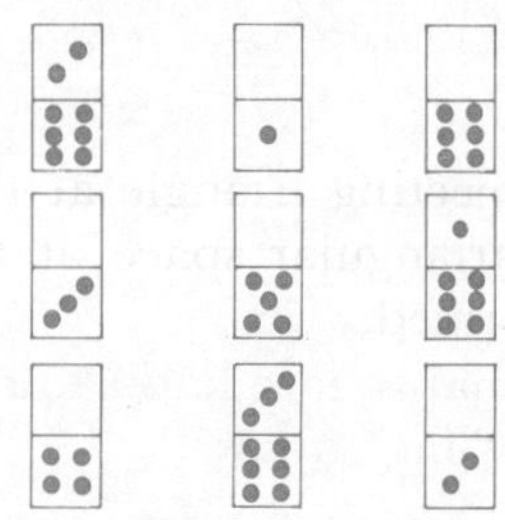

* • Place the full set of dominoes in a square so that the sum of the dots on each side is 44.
• These dominoes form a "magic square". What is the "magic constant"?

* • Form another "magic square", with a constant of 21, using these tiles:

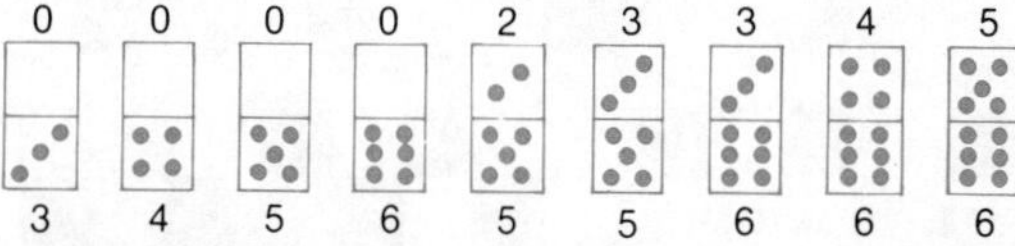

* Symmetry strategy for two players

Two players take turns at placing a domino anywhere on a rectangular board with a total area equal to the total area of the full set of dominoes. What are some possibilities?

Each domino must be put down flat and no other piece may be moved. The player who puts down the last domino wins.
Try to discover a way that the first player can always win.

Chess

✱ How many squares on a chessboard?
Suppose every square of a 5 × 5 chessboard is occupied by a knight. Is it possible for all twenty-five knights to move simultaneously in such a way that at the end *all* the squares are occupied as before?

✱ **Chessboard paradox (The principle of concealed distribution.)**
(A paradox is something that seems to be correct but it contradicts our common sense.)
On 1 cm grid paper make a miniature chessboard then cut it along the oblique line, as shown in Diagram 1. Part B is then moved downward (Diagram 2).

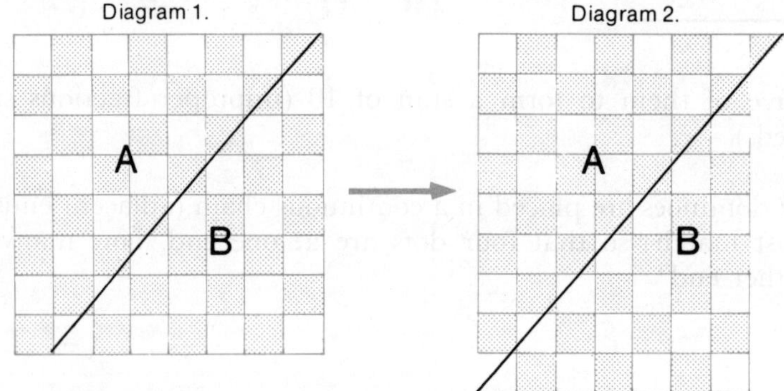

If the projecting triangle at the top right-hand corner is snipped off and fitted into the triangular space at the lower left corner, a rectangle of 7 × 9 units will be formed.
The original square had 64 units, but now there are only 63! What has happened to the missing square?

A crazy chess player cut up his cardboard chessboard into fourteen parts.

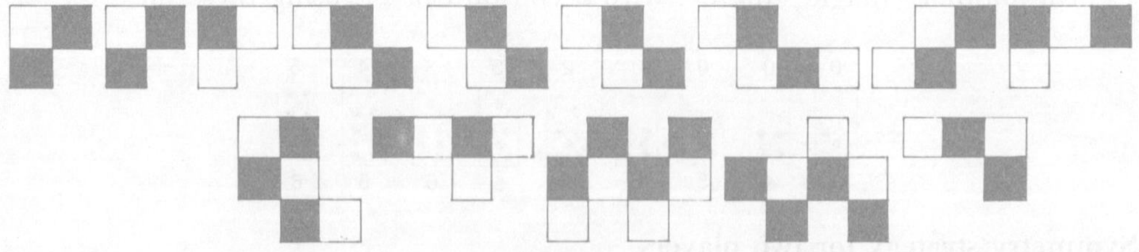

Make a set of chessboard pieces using 1 cm grid paper then see if you can arrange them correctly back into their original form.

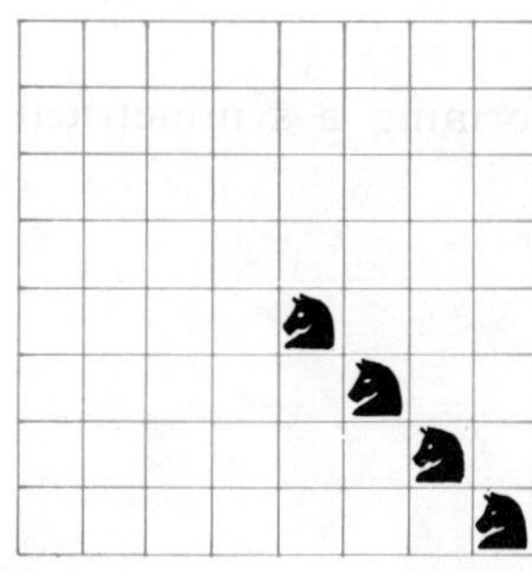

✱ Divide the chessboard into four congruent pieces, each with a knight on it.

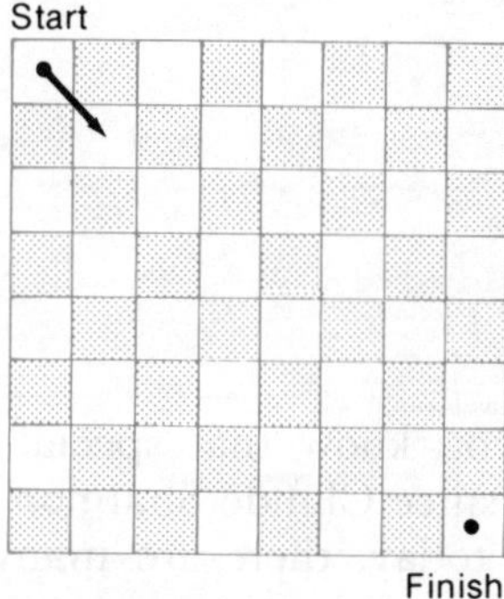

✱ With just seventeen straight lines go through all the white squares (some several times) without crossing any black square en route.

An Ancient Recreation

If a knight is placed on a square of a chessboard and a continuous path of the re-entrant knight's moves is drawn so that it visits each of the sixty-four squares once only, the result is a closed tour. Many interesting, symmetrical patterns can be created.

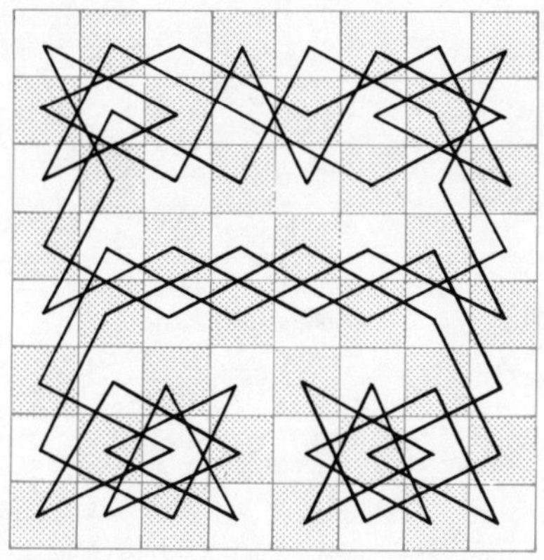

Paths with exact fourfold symmetry (unchanged by 90° rotation) are not possible on an 8 × 8 board although five such patterns are possible on a 6 × 6 board.

ACTIVITY

Trace a re-entrant knight's tour on this simple board, forming a symmetrical pattern. Use grid paper.

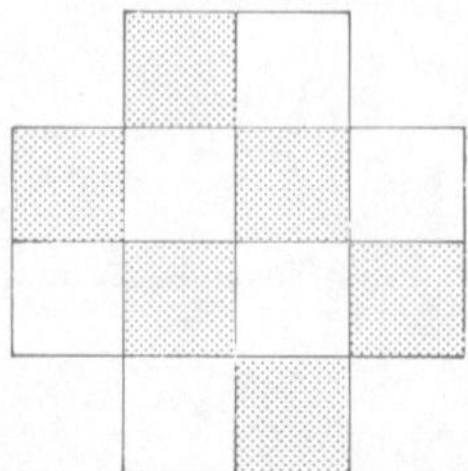

Did you know that special computers can play chess? It is over twenty-seven years since Claude Shannon programmed a machine to play a satisfactory game. Even today, there are many grand masters in the world who can outplay the computers!

Cubes

Assemble these cubes on cardboard, numbering the sides as shown. Then arrange them so that along each side you have the digits 1, 2, 3 and 4.

	3		
3	4	2	1
	3		

	1		
2	1	3	4
	2		

	4		
1	2	4	2
	3		

	3		
2	1	1	3
	4		

Unit 4 Pascal's Triangle

The Frenchman Blaise Pascal (1623–1662) was exceptional, even when he was a young boy. When he was sixteen he discovered the theorem now known as **Pascal's theorem** and wrote an "Essay on Conics" in which he deduced over 400 propositions in geometry, as special cases of his theorem.

At the age of eighteen, he invented and made the first calculating machine, (a forerunner of today's modern desk calculator), a replica of which is in the Science Museum in London.

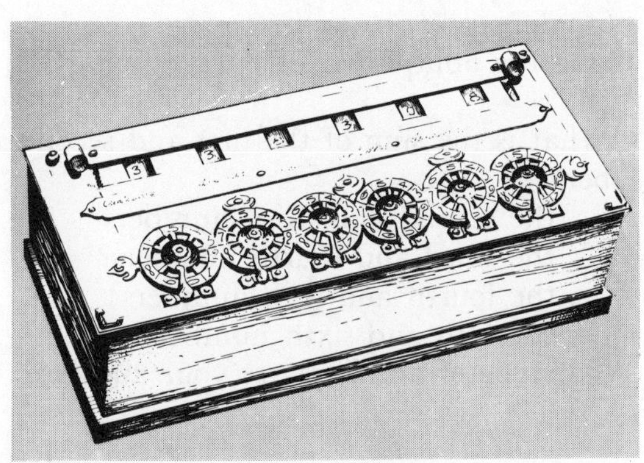

He studied the geometry of the cycloid as a result of suffering from a toothache! He needed something to take his mind off the pain, and as a result made many discoveries about this "curve of a rolling wheel".

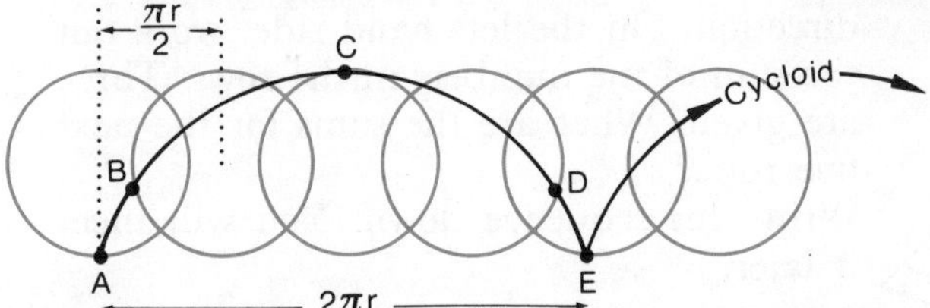

The position of each circle is shown after each quarter turn.

In conjunction with Fermat, Pascal later made many important contributions to the theory of chance and probability. At the age of thirty-one he abandoned mathematics altogether for theological and moral studies. Today his portrait appears on some French stamps.

Pascal's Triangle

This is a pattern of numbers from which binomial probabilities can be easily determined. Each row of the triangle is formed by adding the pair of numbers directly above it, to the left and right. It was known to Omar Khayyam about 1100 A.D. and was published in China about 1300 A.D. However, it is generally known as Pascal's triangle because of the amount of work he did on it.

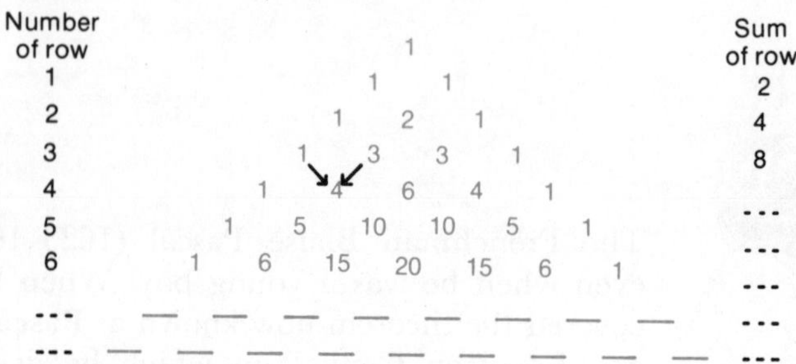

Copy these six rows of Pascal's triangle then complete two more rows, with the row numbers on the left and the sum of the rows on the right.
Underneath, write down the number series made by each diagonal line, commencing with 1, 1, 1, 1 . . .; 1, 2, 3, 4 . . . .
By multiplying, find the value of 11^2, 11^3 and 11^4. Are your answers in the triangle?
Using the oblique set of numbers marked, answer these questions:

- What is the sum of the first and second numbers?
- . . . the second and third numbers?
- . . . the third and fourth numbers?
- . . . the fourth and fifth numbers?
- . . . the fifth and sixth numbers?

What type of numbers are your answers?

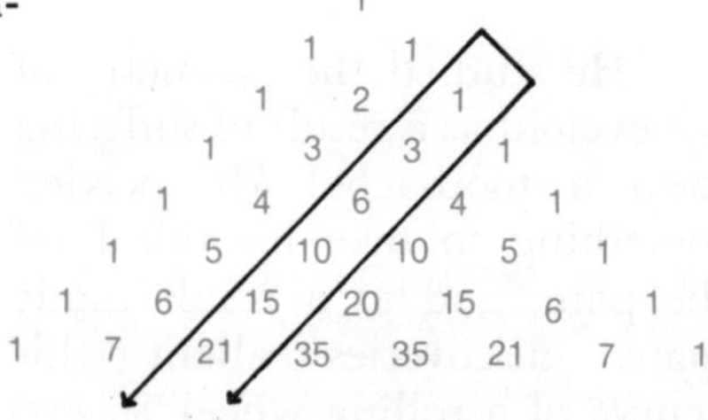

Here is a series of rows tilted in another direction. On the left hand side, work out the sum of the numbers in the rows. Three are given. What are the sums for the next five rows?
Write this sequence down. You will meet it later.

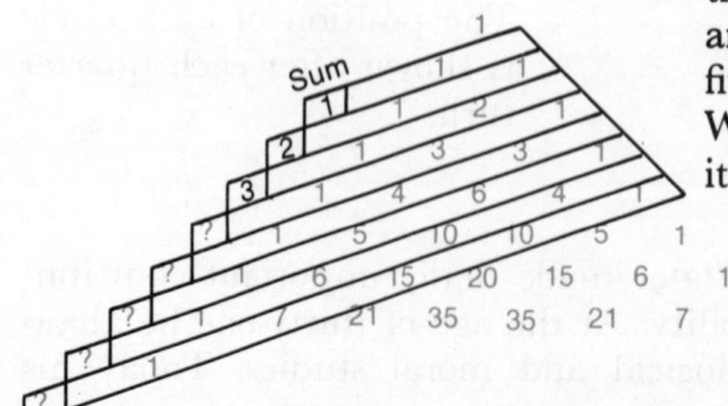

This is a diagram from a Chinese book written around 1303. Do you recognise it?

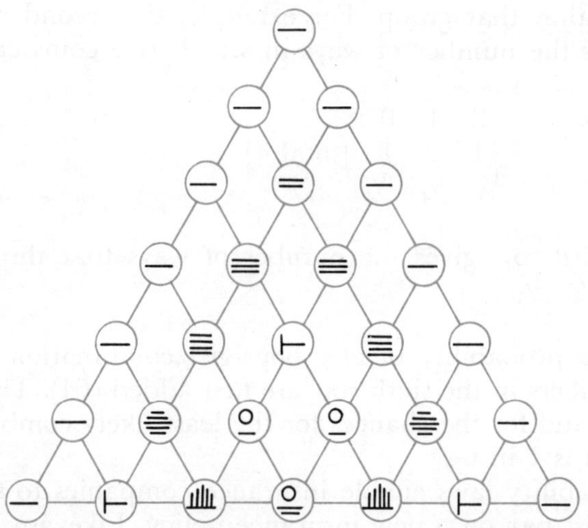

What is the symbol for 6? 10? 15? 20?

Using Pascal's method of calculating rows, complete this triangle.

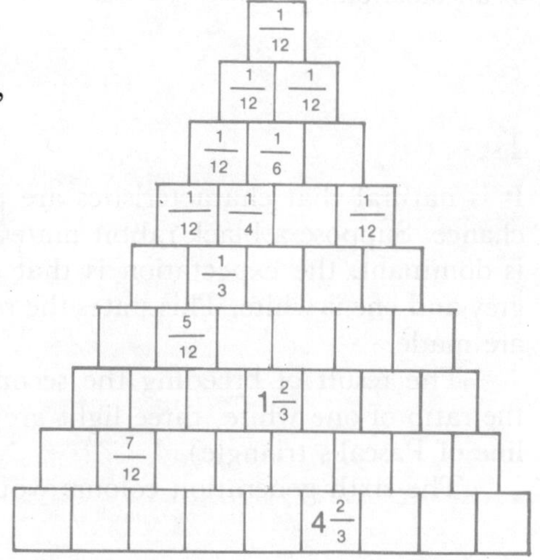

Each row of Pascal's triangle should be compared with the expansions of $(x + y)^n$, which you will use in your senior work.

$(x + y)^0$	1	
$(x + y)^1$	$1x^1 + 1y^1$	
$(x + y)^2$	$1x^2 + 2xy + 1y^2$	Note the
$(x + y)^3$	$1x^3 + 3x^2y + 3xy^2 + 1y^3$	co-efficients
$(x + y)^4$	$1x^4 + 4x^3y + 6x^2y^2 + 4xy^3 + 1y^4$	
$(x + y)^5$	$1x^5 + 5x^4y + 10x^3y^2 + 10x^2y^3 + 5xy^4 + 1y^5$	

Pascal's triangle is used extensively in the subject of *probability.* Instead of using probability trees, this triangle is a ready reference for finding the odds governing combinations. Computations are always made horizontally.

The sum of the numbers in any row gives the total arrangements of combinations possible, within that group. For example, the second row of the triangle (1 2 1) represents the number of ways in which two coins can land when they are tossed.

Number of heads.	2	1	0	
Number of ways.	1	2	1	[total 4]
Probability.	$\frac{1}{4}$	$\frac{2}{4}$	$\frac{1}{4}$	

Similarly, the third row gives the number of ways that three coins may land, that is, $\frac{1}{8}$ $\frac{3}{8}$ $\frac{3}{8}$ $\frac{1}{8}$.

To determine the probability of any boy–girl combination in a family of six children, the numbers in the sixth row are first added (64). The numbers at each end of this row stand for the chances for the least likely combinations – all boys or all girls – and is 1 in 64.

Today, probability laws enable insurance companies to set the rate a fifty-year old man must pay on a new insurance policy. Likewise, they help political reporters to determine, from any given sampling of voters, the probable result of an election.

Heredity

It is natural that characteristics are passed on to offspring by their parents, by chance. Suppose a black rabbit mates with a white rabbit. If neither colour gene is dominant, the expectation is that of every four young, one is black, two are grey and one is white. This states the ratio to be expected if thousands of breedings are made.

The result of breeding the second generation together produces rabbits in the ratio of one white, three light grey, three dark grey and one black (the third line of Pascal's triangle).

The sixth generation colours would tend to be in the ratio of:

1	one white
6	six very pale grey
15	fifteen pale grey
20	twenty medium grey
15	fifteen dark grey
6	six very dark grey
1	one black.

Notice that out of the sixty-four cases, only one is in each of the two original colours while no fewer than fifty of them are in the middle colour range. This grouping (with a large majority of cases clustered around the average) is characteristic of all inherited qualities. It makes mass production of clothes possible as most of the population is able to be fitted from quite a small range of sizes.

Coin Tossing

The expectation of heads and tails after a large number of trials is given by the line of the triangle corresponding to the number of coins being tossed at each trial.

When a single coin is tossed many times, the expected ratio of heads to tails is given by the co-efficients of the expansion of $(x + y)^1 = x + y$, that is, one head to one tail.

The result after tossing two coins many times is given by the co-efficients of $(x + y)^2$ or 1 2 1, while that of six coins can be found by referring to the sixth line, giving 1 6 15 20 15 6 1. This means six heads (or tails) could be expected once, five heads and one tail (or vice versa) six times, four heads and two tails (or vice versa) fifteen times and three of each kind, twenty times in every sixty-four trials, on the average.

Use Pascal's triangle to work out the probability of this problem.

If a coin is tossed ten times, what is the probability that it will come up heads every time? (This question is equivalent to that of finding the probability of getting ten heads when ten coins are tossed.)

Pascal's Theorem states that if a hexagon is drawn inside a conic section, the points of intersection of its three pairs of sides, produced, will lie on a straight line.

The hexagon *ABCDEF* is drawn inside the envelope of the "ellipse", constructed from four rectangular parabolae.

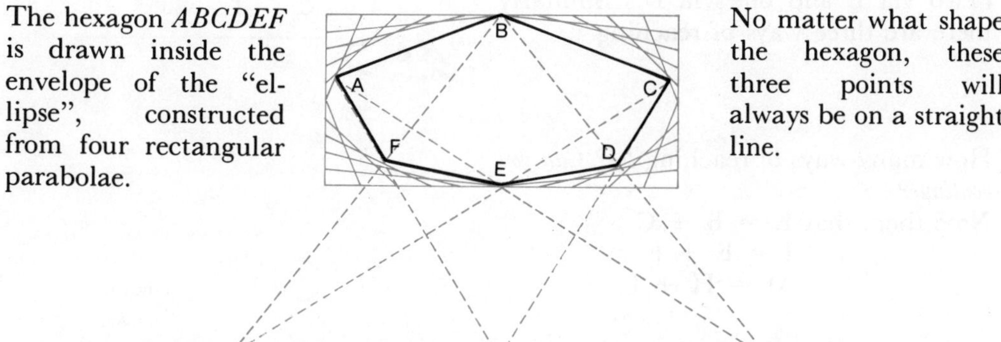

No matter what shape the hexagon, these three points will always be on a straight line.

Just for Fun

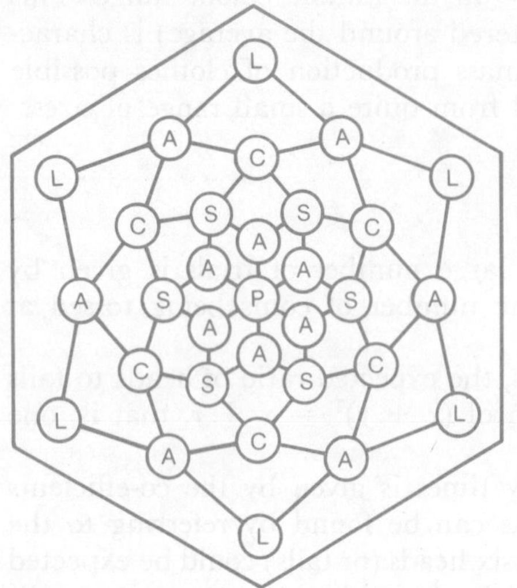

In how many ways can you spell **Pascal** in this figure, starting from the centre P and working out towards the edges?

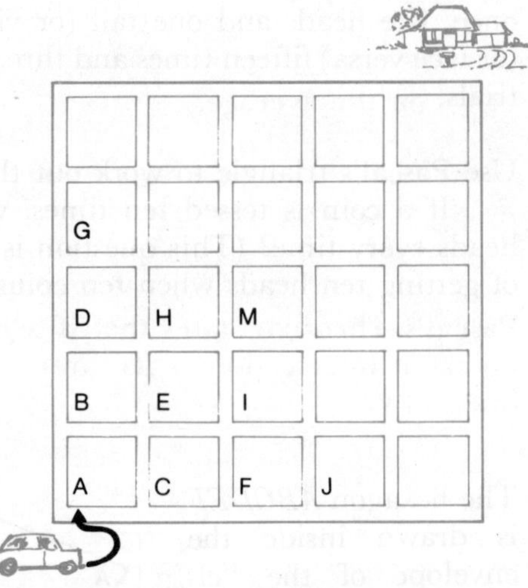

Show me the way to go home

In how many different ways is it possible for the motorist to go from A to his home Z by the shortest route?

In how many ways is he able to go to C, directly? to B?

This means he has two ways of reaching E (via B or C).

From B there is only one route to D. Thus he has three ways of reaching H. (Two via E and one via D.) Similarly there are three ways of reaching I.

How many ways of reaching M? *Can you continue?*

Note then, that E = B + C

I = E + F

M = H + I

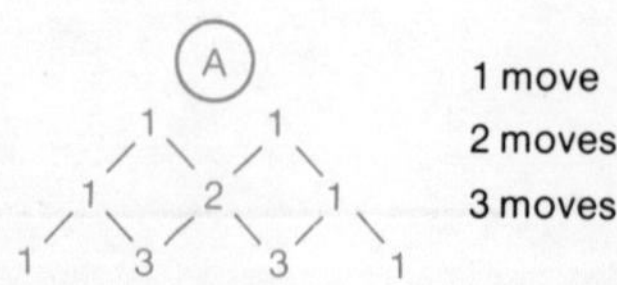

If you write the number of possible ways in each intersection you will recognise the pattern. Continue this until you find Z's number.

Unit 5 Fibonacci Numbers

This sequence of numbers owes its name to the great Italian mathematician Leonardo da Pisa (often called Fibonacci, a contraction from "Filius Bonacci" meaning son of Bonacci) who was born in Pisa between 1170 and 1175, and lived until 1230. The construction of the famous Leaning Tower was begun during his lifetime, although it was not completed for nearly two centuries.

He was educated at Bugia and travelled about the Mediterranean collecting information about mathematics. In 1202 he returned to Pisa and published *Liber Abaci*, a book which established the introduction of the Arabic notation in Europe and provided a foundation for future development in arithmetic and algebra. It discussed such topics as the basic operations (multiplication, addition, subtraction and division), fractions, prices of goods, bartering, problem solving, the square root and also mentioned various methods of counting – Arabic, the abacus and finger reckoning.

In 1220 he published another book on geometry called *Practica Geometria.*

Then in 1225 the Holy Roman Emperor Frederick II came to Pisa with a group of mathematicians to publicly test Fibonacci's amazing skill at problem-solving. One problem, presented at the contest was "To find a square which remains a square if it is decreased by 5 or increased by 5". Fibonacci was able to write down the fractional answer in a relatively short time.

What Are Fibonacci Numbers?

This sequence is found by finding the sum of two consecutive terms to give the next term. Thus $1 + 1 = 2$, $1 + 2 = 3$, $2 + 3 = 5$ and so on.

$$1 \quad 1 \quad 2 \quad 3 \quad 5 \quad 8 \quad 13 \quad \ldots$$

If two consecutive terms in the series are y and x, then

$$x^2 - xy - y^2 = 1 \text{ or } x^2 - xy - y^2 = -1.$$

Test this by substitution.

EXERCISES

1. Write down the first fifteen terms of the sequence. Remember that the first two terms are each 1; the fifteenth term is 610.
2. What type of number is every third one?
3. Is every fourth number divisible by 3, every fifth by 5 and the fifteenth by 10?
4. Find the sum of the first *five* terms and compare your answer with the eighth term.

5. Add the first *twelve* Fibonacci numbers and look for another number that is related to it.
6. Can you guess the sum of the first thirteen terms of the sequence without adding them?

There are four **next-door** Fibonacci numbers.

2	3	5	8

Multiply the two outside numbers $2 \times 8 = 16$.
Multiply the two middle numbers $3 \times 5 = 15$.
What is their **difference**?

7. Try this procedure with other sets of four next-door terms.

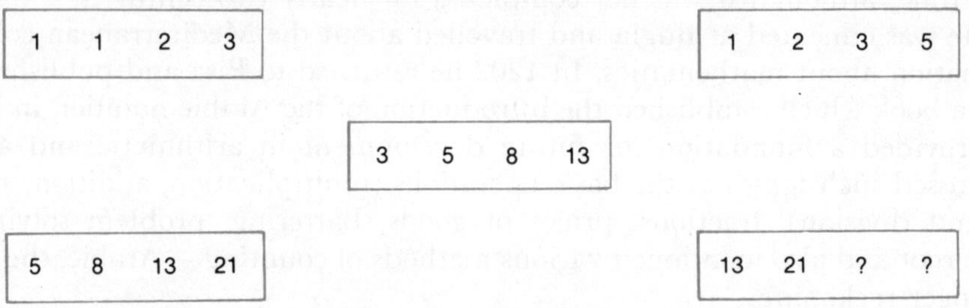

Do you always get 1?

8. Would it be possible to construct a triangle with sides that are three Fibonacci numbers?
9. Cubing Fibonacci numbers.
   $8^3 = 512$
   $5^3 = 125$
   $3^3 = 27$
   $512 + 125 - 27 = 610$ (another one!)
   Will this always occur?
10. The product of the *first* and *last* terms of any three consecutive Fibonacci numbers is always one less than the square of the middle one.

Example: 8 13 21 ⟶ $8 \times 21 = 168$
$13^2 = 169$

168 is 1 < 169.

Try four more examples for yourself.

11. Another famous 18th century French mathematician named Joseph Lagrange discovered a pattern in the *remainders* formed by dividing each term of the Fibonacci sequence by 4.

    It begins with the first 4 terms being repeated.

    1 1 2 3 1 0 . . . .

    Find the pattern of remainders by working out the first twelve numbers.

12. Find the squares of the first eight terms of the sequence, commencing

1 1 4 9 . . . .

Now add each pair of consecutive squares to make a new sequence. What do you notice about it?

13. Here is another interesting pattern based on squares.

$$1^2 + 1^2 = 1 \times 2$$
$$1^2 + 1^2 + 2^2 = 2 \times 3$$
$$1^2 + 1^2 + 2^2 + 3^2 = 3 \times 5$$
$$1^2 + 1^2 + 2^2 + 3^2 + 5^2 = 5 \times 8$$

Work out the next two lines of the pattern and check to see if they are correct. This illustrates that the sum of the squares of the first n numbers of a Fibonacci series is the product of two adjacent numbers.

14. The square of each number, reduced by the product of the preceding and following numbers is alternately +1 and −1.
Check this rule by calculating . . .

(Remember your order of operations.)

$$2^2 - 1 \times 3 =$$
$$3^2 - 2 \times 5 =$$
$$5^2 - 3 \times 8 =$$
$$8^2 - 5 \times 13 =$$

15. Cubing terms.

Copy down and work out this pattern.

$$2^3 + 3^3 - 1^3 =$$
$$3^3 + 5^3 - 2^3 =$$
$$5^3 + 8^3 - 3^3 =$$

Can you add two more lines?
Do you recognise the type of answers?

16. Taking ratios of consecutive terms in the sequence, we have:

$$1, \frac{2}{1}, \frac{3}{2}, \frac{5}{3}, \frac{8}{5}, \frac{13}{8}, \frac{21}{13}, \frac{34}{21}, \frac{55}{34}, \_, \_, \_.$$

These converge to 1·6183, which is called the *Golden Ratio.* (See Unit 8.)

If the fractions are written as decimals, in order, they are: 1·0, 2·0, 1·5, 1·66, 1·6, 1·625, 1·6154 . . ., 1·61905 . . ., 1·61765 . . ., 1·61818 . . ., 1·61798 . . ., 1·618055 . . ., 1·618025 . . ., 1·618037 . . . approaching the golden ratio which plays a part in design and architecture.

This is Pascal's triangle, dealt with in Unit 4.
Find the sum of the numbers on each dotted line. What do your answers form?

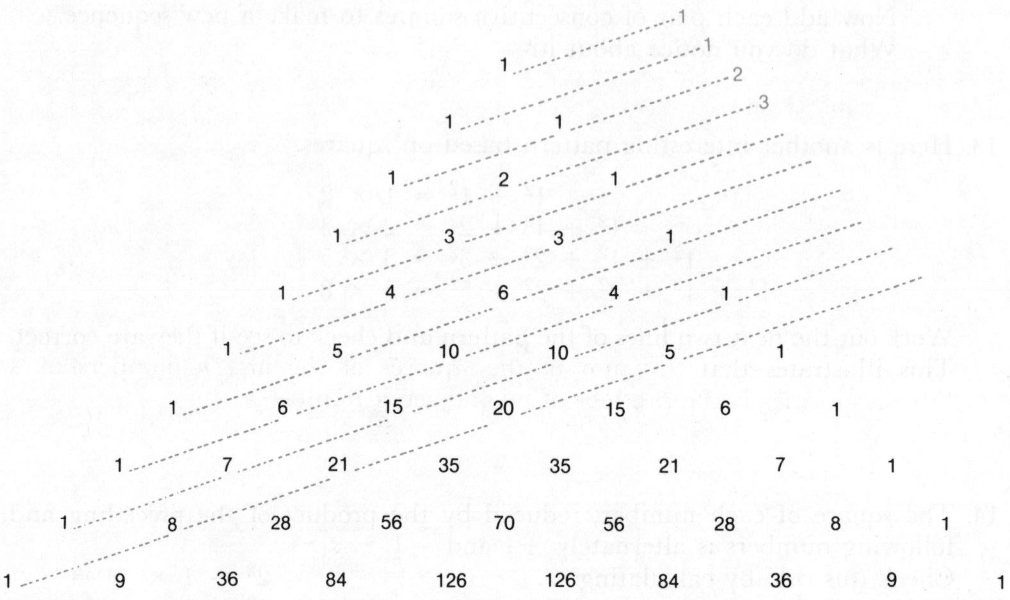

Fibonacci Sequence Tricks

1. Impress your friends with mental arithmetic.
   1 1 2 3 5 8 13 21 34 55 89 144 233 377 610 987 1597 2584.
   Tell your friend to cut off the sequence at any number and you will add up the preceding numbers (and the cut off number) mentally!
   If 144 is the cut off, your answer is 376.
   If 610 is the cut off point, you say the answer is 1596.
   Can you see how to obtain your answers?
2. This trick involves adding, almost instantly, any ten numbers of a Fibonacci sequence.

   Ask a friend to write down any two Fibonacci numbers, one below the other, and to add them together to form a third line. Add this line to the one above, to make a fourth line and so on, until there are ten numbers in a vertical column.

   The "performer" keeps his back turned the whole time. After the ten numbers are written the "performer" turns around, draws a line below the column and quickly writes the total. The secret is to multiply the fourth number from the bottom by 11. *Try it.*

Fibonacci devised some problems in his books. He posed this one in *Liber Abaci.*

A pair of rabbits one month old are too young to produce more rabbits, but suppose that in their second month and every month thereafter they produce a new pair. If each new pair of rabbits does the same, and none of the rabbits dies, how many pairs of rabbits will there be at the beginning of each month?

Another one says:

Suppose a cow which first calved when in her second year brings forth a female calf every year and each she-calf, like her mother, will start calving in her second year and will bring forth a female calf every year – how many calves will have sprung from the original cow and all her descendents in twenty-five years? (*Answer:* 121 392.)

Both the solutions rely on Fibonacci numbers!

Applications

The Fibonacci sequence appears in such unrelated topics as the family tree of the male bee, the keyboard of a piano and frequently in biology.

A drone bee (male) has a mother but no father, as the queen's unfertilised eggs hatch into drones. The queen's fertilised eggs produce either worker bees or queens. The diagram shows why the number of ancestors of a drone must in any generation be a Fibonacci number.

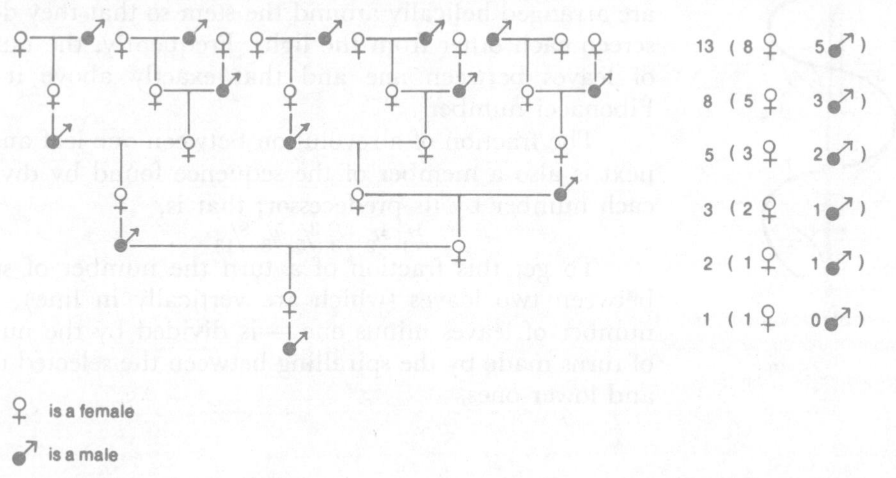

Look at the similarity between this example and an octave on a piano keyboard.

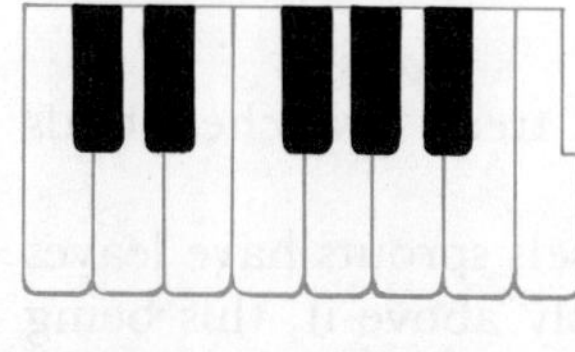

The thirteen elements correspond to thirteen semi-tones (8 white, 5 black) of the chromatic scale.

Botanical Applications

Plant Growth (*Phyllotaxis* or leaf arrangement)

This Fibonacci sequence is also demonstrated in the number of petals of many species of flowers. Of course, we cannot expect to find these exact numbers on *every* plant, owing to various mutations that may occur, but in general a mean (average) score would tend towards these figures.

Examples:		
	2 petals	uncommon
	3 petals	lilies and irises
	5 petals	buttercups, some delphiniums (This is the most common group.)
	8 petals	some delphiniums, but not as common
	13 petals	marigolds
	21 petals	some asters
	34, 55 and 89 petals	daisies – also common

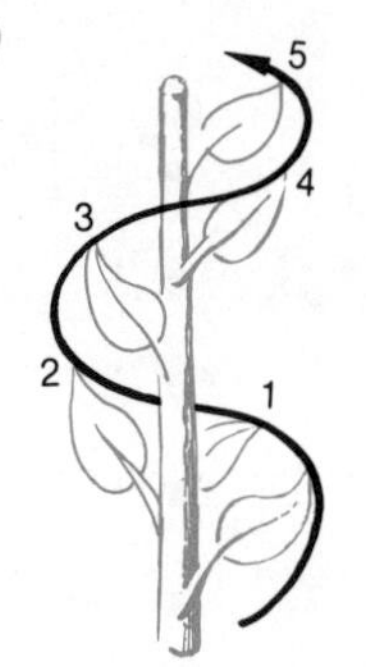

Number of revolutions = m
Number of leaves = n
Arrangement = $\frac{m}{n}$ spiral

Fibonacci numbers also appear in the arrangement of leaves on the stems of plants. On many plants the leaves are arranged helically around the stem so that they do not screen each other from the light. Frequently, the number of leaves between one and that exactly above it is a Fibonacci number.

The fraction of a revolution between one leaf and the next is also a member of the sequence found by dividing each number by its predecessor; that is,

$$1/1 \quad 1/2 \quad 2/3 \quad 3/5 \quad 5/8 \quad 8/13 \ldots$$

To get this fraction of a turn the number of spaces between two leaves (which are vertically in line) – the number of leaves minus one – is divided by the number of turns made by the spiralling between the selected upper and lower ones.

Different plants have characteristic angles of divergence of adjacent leaves. The angle is expressed as a fraction of 360°. Here are common examples:

elm $1/2$ oak or cherry $2/5$
beech $1/3$ poplar or pear $3/8$
willow $5/13$

The same angle is preserved in the arrangement of each tree's branches, buds and flowers.

Examples of the 1 : 2 ratio abound: broccoli and brussels sprouts have leaves that spiral five times beween one leaf and the one vertically above it, this being on the ninth one (8 spaces) and so falls into the $5/8$ category.

A pine cone generally has eight clockwise parts and thirteen anticlockwise parts on the spirals.

This falls into the $^{8}/_{13}$ class, as does the pineapple, with its spiral whorls made of roughly hexagonal cells.

Here is a diagram that illustrates the arrangement of cells. If the pineapple's surface is regarded as a cylinder, cut and spread out on a plane, the pattern of spirals becomes obvious. Follow the number patterns obliquely upwards.

In a sunflower there are two opposite sets of rotating spirals formed by the arrangement of the individual florets in the head. There are twenty-one spirals in a clockwise direction and thirty-four in the anticlockwise direction. This 21 : 34 ratio corresponds to two adjacent terms in the sequence.

The Fibonacci numbers are, therefore, a set of whole numbers which satisfy almost exactly an exponential law of growth.

They form, however, a different type of sequence from that which relates to the growth patterns found in the leaves, branches and petals of plants, which are based on whole numbers. They can be shown to be connected to another type of growth in the biological field, continuous growth, such as height, weight and velocity in organic bodies.

Graphically, the Fibonacci numbers form a continuous curve that is the graph of the function

$$y = 0{\cdot}4472 \times (1{\cdot}61803)^x$$

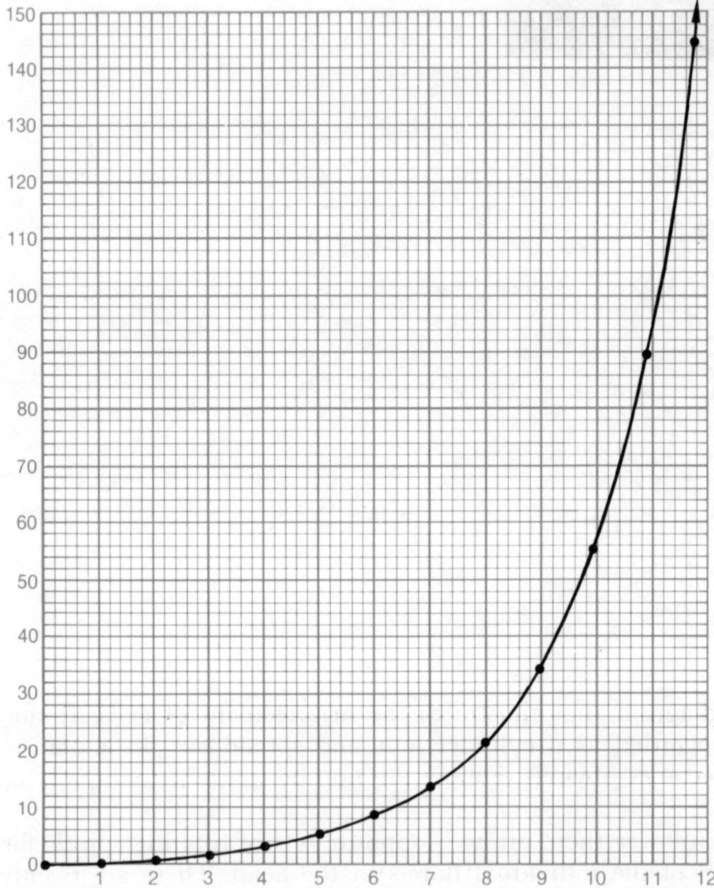

The sequence of Fibonacci numbers is also closely related to the logarithmic spiral.

EXERCISE

On 2 mm grid paper draw a rectangle 89 units long by 55 units wide. (Diagram 1.)

Now draw line segments in the rectangle to divide it into a series of squares with sides of 55, 34, 21, 13, 8, 5, 3, 2, 1 and 1 respectively (Diagram 2). Then draw an arc that is a quarter circle from a corner of each square, starting with the largest square, to form a continuous curve. The diagram shows the first two arcs drawn in.

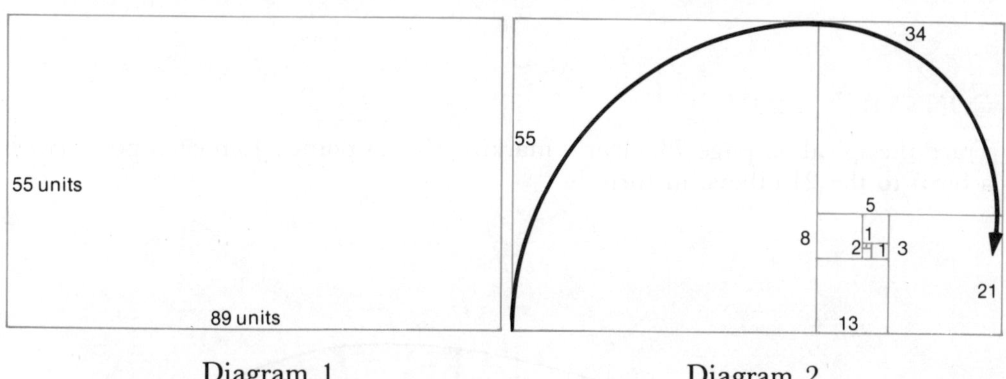

Diagram 1 Diagram 2

The resulting curve, based on the Fibonacci sequence, is very similar to the shape of the chambered nautilus.

The Chambered Nautilus

This is a magnificent shell which is the silent toil of a mollusc which is seldom seen alive. As it grows, the animal secretes partitions within its expanding pearly shell, creating a series of ever-larger rooms. By varying the gas content of the

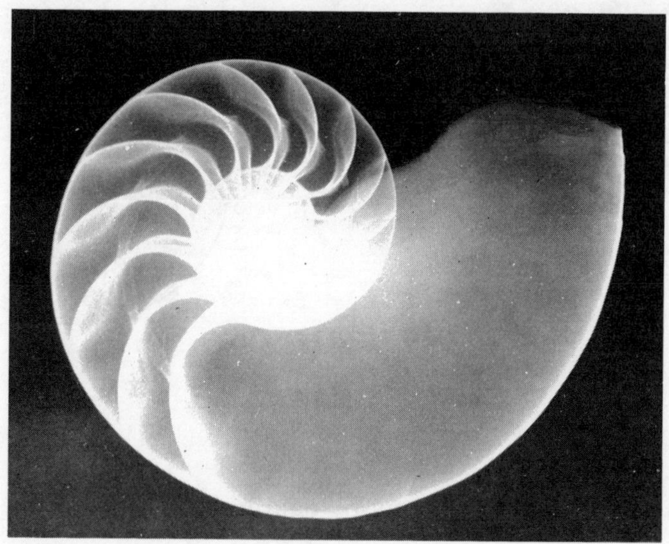

abandoned chambers, it changes its buoyancy and dives as deep as 600 m at night. Through a tube connected to its body, the nautilus fills the shell chambers with gases so that it will rise in the water and then reabsorbs the gases to sink. Using its muscular funnel, it jets horizontally while feeding with its many tentacles near coral reefs during the day. A cephalopod ("head-foot"), the most highly developed of molluscs, the once-abundant chambered nautilus stopped evolving eons ago. Today, the western Pacific harbours the last four species of this living fossil.

CONSTRUCTION

Trace the spiral on page 73, clearly marking the 25 points. Join each point (with a biro) to the 24 others, in turn.

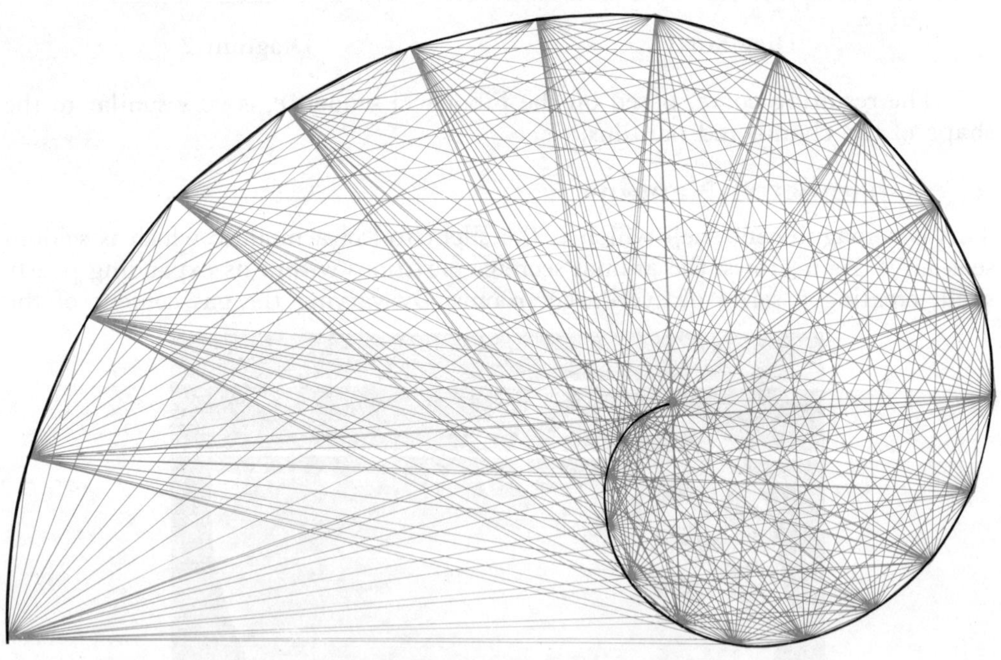

Your finished result should be a "mystic nautilus".

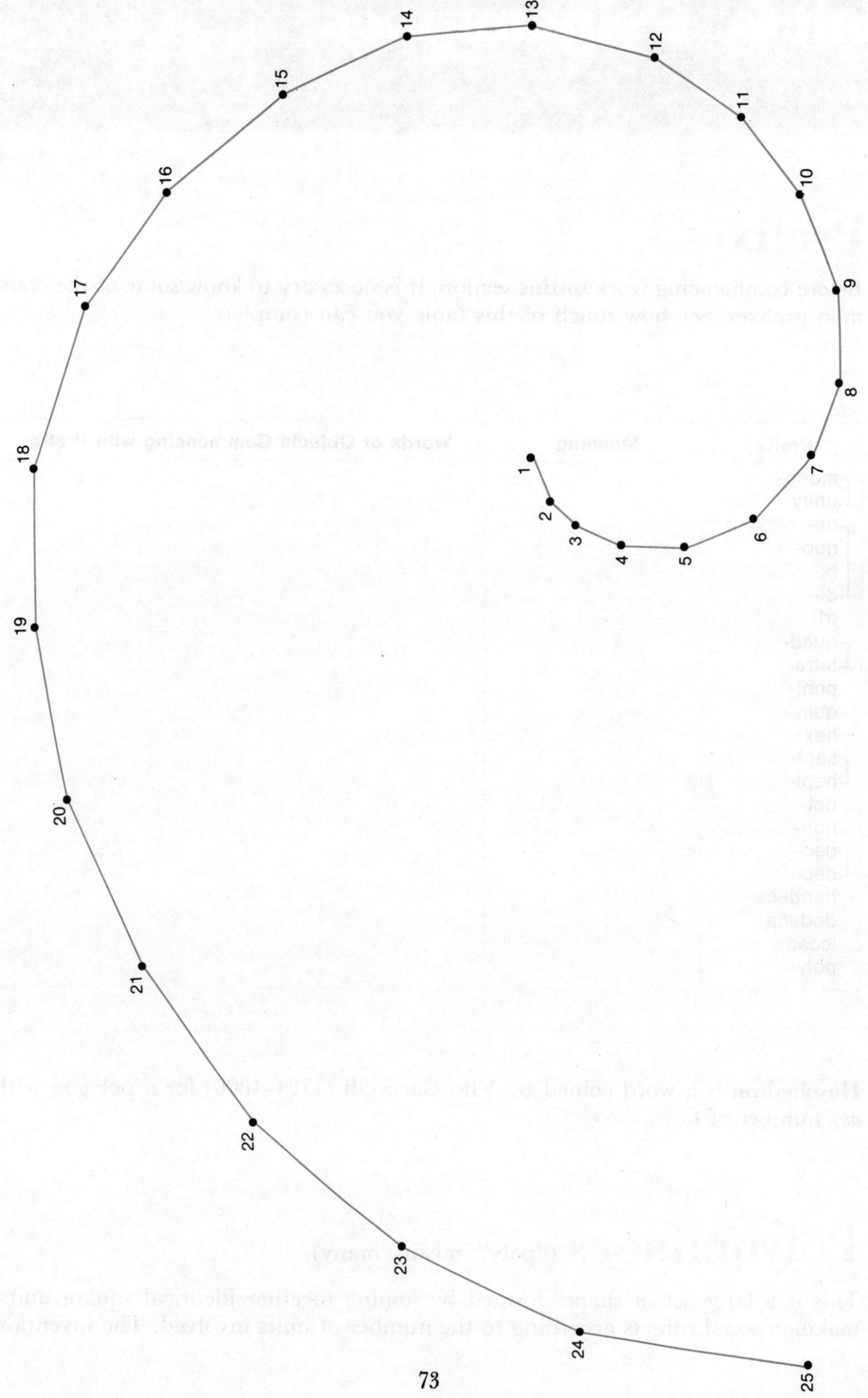
1
2
3
4
5
6
7
8
9
10
11
12
13
14
15
16
17
18
19
20
21
22
23
24
25

Unit 6 Polyominoes

Prefixes

Before commencing work in this section, it is necessary to know some of the common prefixes. See how much of this table you can complete.

Prefix	Meaning	Words or Objects Commencing with Prefix
mono- unity		
do- duo- bi- di-		
tri-		
quad- tetra-		
pent-		
quin-		
hex-		
sept- hept-		
oct-		
non-		
dec- deci-		
hendeca-		
dodeca-		
icosa-		
poly-		

Hosohedron is a word coined by Vito Caravelli (1724–1800) for a polygon with *any* number of faces.

Polyominoes **("poly" means many)**

This is a large set of shapes formed by joining together identical square units, making special subsets according to the number of units involved. The invention of polyominoes is credited to Solomon Golomb (an American mathematician), who first introduced them in 1953.

The most frequently used ones are:

Monomino

Pentomino

Domino

Hexomino

Tromino (or tr'omino)

Tetromino

Heptomino

What would you call an eight-unit shape?

Monominoes and Dominoes

There is only one way in which a monomino or domino can be drawn. (Rotation does not make a different domino.) They both obviously tile a plane.

Trominoes

There are two ways of drawing trominoes.

A straight tromino

A right tromino

Would these tile a plane?

Note that they are joined by their edges, not vertices alone.
These two methods are *not* allowed:

Tetrominoes

In your exercise book, carefully draw and label the five possible ways of joining four identical squares together. You will have straight, square and skew shapes.

Grid paper makes the exercise a lot faster!

What is the total area of the five tetrominoes?

It is impossible to form these shapes into a 5 × 4 rectangle, but with the addition of a 5-square shape (called a pentomino), a 5 × 5 square may be formed.

Here are two possibilities.

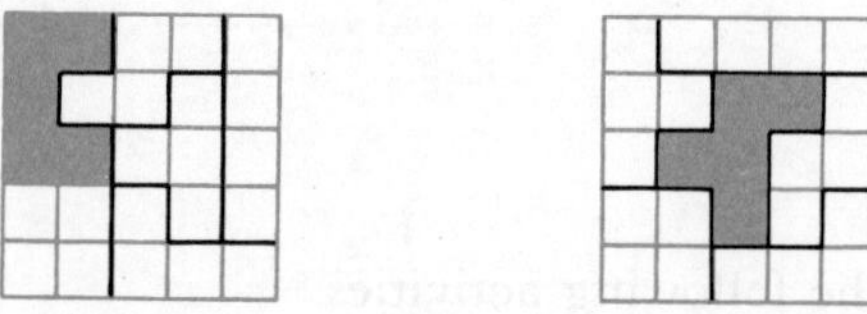

ACTIVITIES

1. Here is a 5 × 5 board with five monominoes coloured in.
   Example: Select one of each of the five tetrominoes and place them on the blank squares. Draw your solution on 5 mm grid paper and paste it in your book.

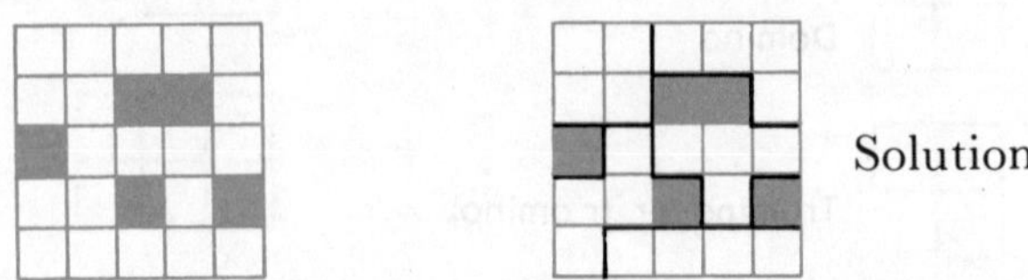

✱ Now try these:

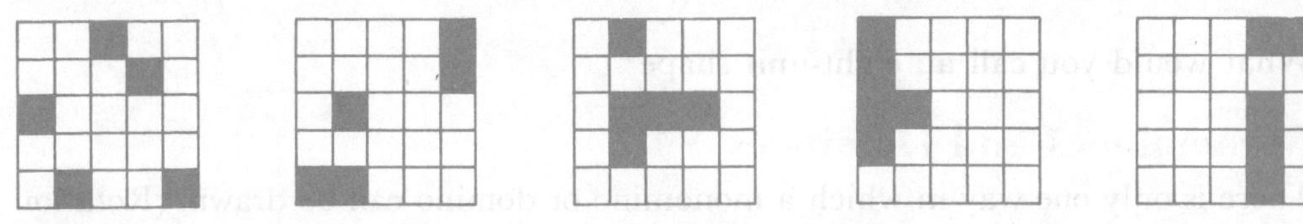

2. **Single shape game.**
   Use six pieces, all this "L" shape, to cover the entire square, except for the monomino in the lower right hand corner.

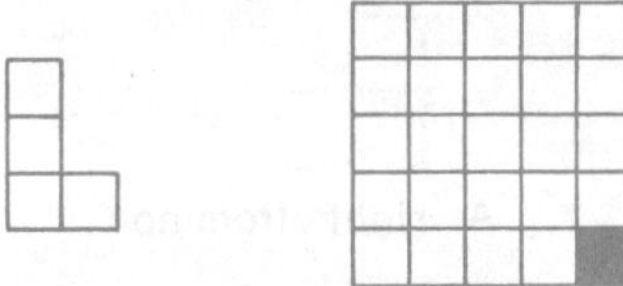

3. On 1 cm grid paper, discover how many of the five tetrominoes will tile the plane. You will probably find that none of them needs to be rotated or reflected. Cut out a sample of each finished work and paste it in your book.

Pentominoes

The word **pentomino** originated in America. It describes a set of twelve shapes, obtained when five regions each one square unit, are joined edge to edge, in all possible ways.

✱ Here are five of them. Copy these, then work out the other seven.

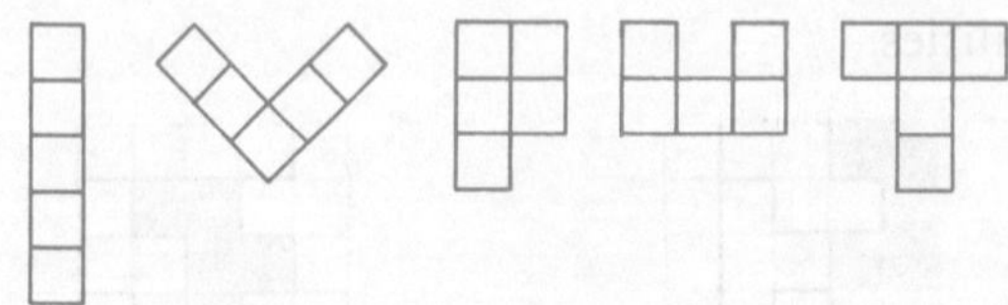

Keep a set for use in the following activities.

Most of them resemble letters of the alphabet . . . I, V, P, U, T (shown here) and W, X, Y, Z, F, L and N.

How many of them are symmetrical (line or rotational)?

ACTIVITIES

* 1. By arranging the twelve pentominoes together, edge to edge, make the largest closed area possible.
2. Combine all the pieces to make the largest number of "holes" possible. Pieces may touch vertex to vertex as well as edge to edge.
3. Use the twelve pentominoes together to cover a page of your grid book.
* 4. By the addition of a 2 × 2 unit shape, make a square with a side of eight units or a rectangle (with an area of sixty-four square units) using all thirteen pieces.
* 5. Form rectangles using the given number of pieces. (Draw your solutions on graph paper.)

5 × 10 rectangle	with only the "Y" pentomino
3 × 5 rectangle	any 3 pieces
4 × 5 rectangle	4 pieces
2 × 10 rectangle	4 pieces
5 × 5 rectangle	5 pieces
3 × 10 rectangle	6 pieces
7 × 5 rectangle	7 pieces
4 × 10 rectangle	8 pieces
9 × 5 rectangle	9 pieces
10 × 5 rectangle	10 pieces
11 × 5 rectangle	11 pieces
2 rectangles, each 5 × 6	12 pieces
10 × 6 rectangle	all the pentominoes

(It is possible to arrange them in such a way that each pentomino touches the border of the rectangle.) Don't give up! There are 2 339 ways of doing this task.

12 × 5 rectangle	all pieces (1 010 ways)
15 × 4 rectangle	all pieces (368 ways)
20 × 3 rectangle	all pieces (2 ways only)

Colour each completed rectangle using no more than four colours.

* 6. Take any of the pieces T, U, Z, W or P and using just four of the remaining pieces, make a duplicate with a factor of 2.
7. Form any desired shape with two pentominoes. Use any other two pieces to form a duplicate or congruent shape.

What is each area? Have they the same perimeters?

Now, using the remaining eight pieces, form the same shape, with dimensions twice as large (a similar figure).

How many square units must the large shape have?

Here is a completed example:

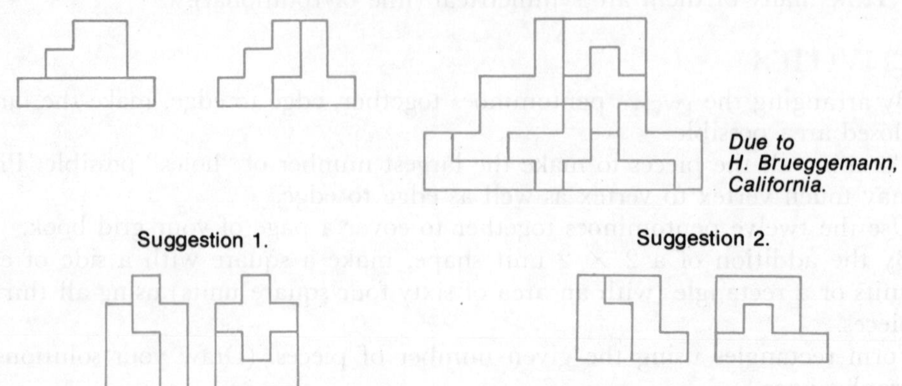

Rearrange the remaining eight pieces to form a larger, similar shape.

✱ 8. Choose any pentomino. Use any nine of those remaining to form a large-scale model of the first shape, which should be three times higher and wider than the small one.

Compare their areas and perimeters.

9. Divide the twelve pentominoes into three groups of four. Find a 20-square shape that each of the groups will make. That is, form three congruent shapes.

One of several solutions is this:

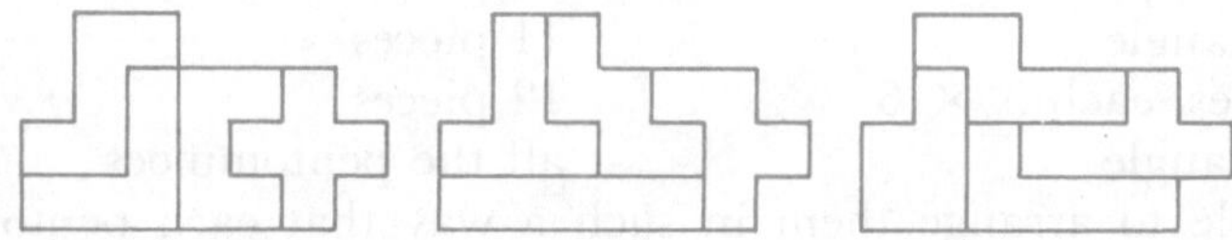

✱ 10.

This cross has four pentominoes in their correct positions. Cover the remaining area with the other eight.

A crossroad is a point where the corners of four pieces meet. How many are there in your solution?

✱ 11. A Game using the Domino

This is a 4 × 4 board of squares with opposite corners cut away. Can you completely cover the board, without overlapping, with domino-shaped tiles? Give a reason for your answer.

A Game using Pentominoes

Using the twelve pentominoes cut to fit the squares on a chess board, players take turns to place a pentomino on the board in any desired position. When a player is unable to do so, the game is finished.

A Class Competition

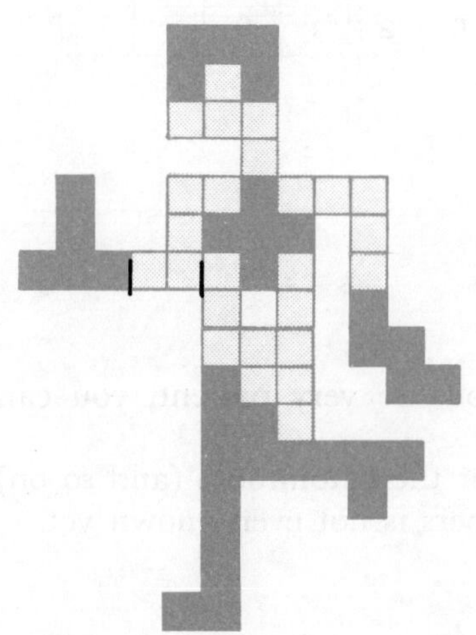

There are many interesting silhouettes that can be made with the twelve pentominoes. A competition could be organised by your teacher to find the most original and imaginative silhouette.

Hexominoes

These are simply six connected squares, joined by edges, to form various shapes. Remember a shape may be turned over, or rotated, but this does *not* represent a new shape.

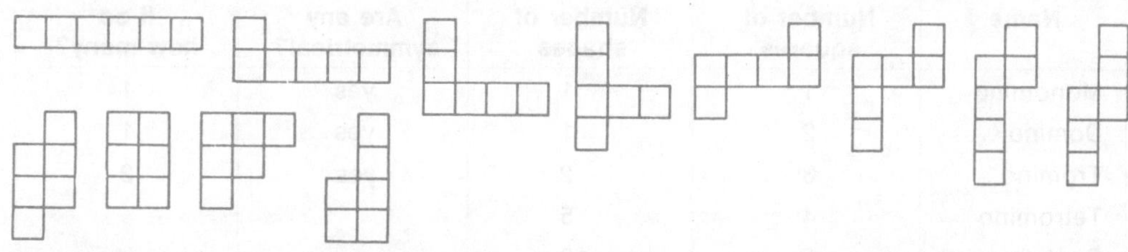

Here are twelve hexominoes. Altogether there are thirty-five of them, all of which tile a plane without reflection. Make a full set in your book, trying to build up similarly related subsets, systematically.

When you have a full set, numbered 1 to 35, you can copy down, continue and complete the following table:

Number	1	2	3	4	5	6	7	8	9	10	11	12	13	14	15
Perimeter in units	14	14	14	14	14	14	14	14	12	10	12	12			
Area in units^2	6	6	6	6	6	6	6	6	6	6	6	6			
Number of axes of symmetry	2	0	0	0	0	0	0	0	0	2	1	0			

Heptominoes

Seven connected squares form a heptomino. If you are very patient, you can build up your own set of 108 different shapes!

Beyond heptominoes, the number of shapes for the octominoes (and so on) is very large. A formula for calculating these numbers is not even known yet.

A Summary

Copy and complete.

Name	Number of squares	Number of shapes	Are any symmetrical?	If so how many?
Monomino	1	1	yes	1
Domino	2	1	yes	1
Tromino	3	2	yes	2
Tetromino	4	5		
Pentomino	5	12		
Hexomino	6	35		
Heptomino	7	108		
Octomino	8	369		

Solving Polyomino Problems by Computer

Golomb once posed the following kind of problem that is something like a jigsaw puzzle.

"Given a number of small polyomino pieces and a large polyomino shape whose area is equal to the sum of the areas of the pieces, find all the ways that the pieces can be placed so as to cover the shape exactly."

Here is one simple example:

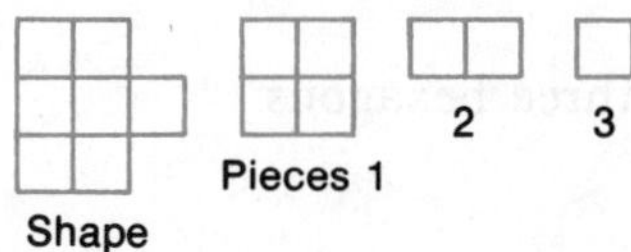

Using 1 cm grid paper show the two possible arrangements.

Allowing the transformations of reflection and rotation and using matrices and combination trees it is possible for a computer to work out much more complex problems in a small fraction of the time a person would take.

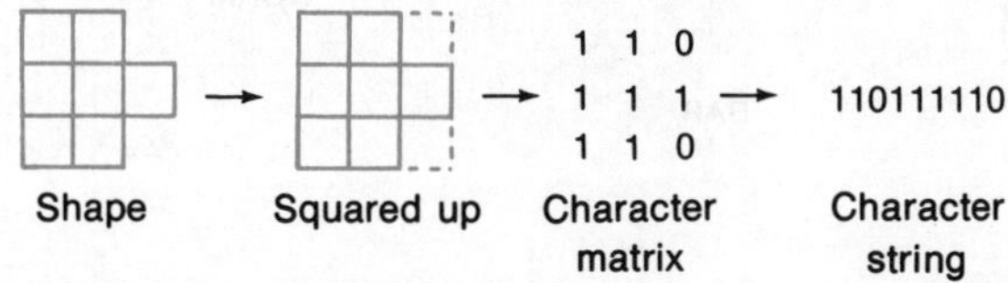

The following problem was solved in six minutes on an ICL 1903T computer. Each unit square was either red, white or blue and the pieces were coloured on both sides (for reflection). There are actually eight solutions, related to the diagonal symmetry of pieces 1 and 4.

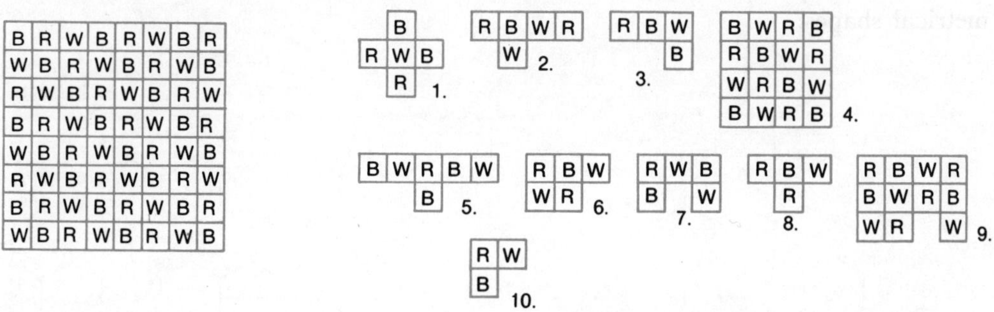

Copy the 8 × 8 square into your exercise book then try to outline the ten pieces, in colour, to fit into the entire square. Perhaps you might like to do this puzzle on graph paper first.

Polyhexes

Regular hexagons, joined edge to edge as in polyominoes, form a large set of polyhexes, with tetrahexes (4) and pentahexes (5) being the most common subsets.
There is only one way to join two hexagons together, edge to edge.

There are three ways to join three hexagons.

Tetrahexes

In the tetrahex group (four hexagons) there are seven distinct shapes.

PISTOL BAR WORM ARCH

BEE PROPELLER WAVE

These shapes resemble the structural diagrams of benzene ring compounds.

✱ Cut out a complete set of the tetrahexes and arrange them to fill in these symmetrical shapes:

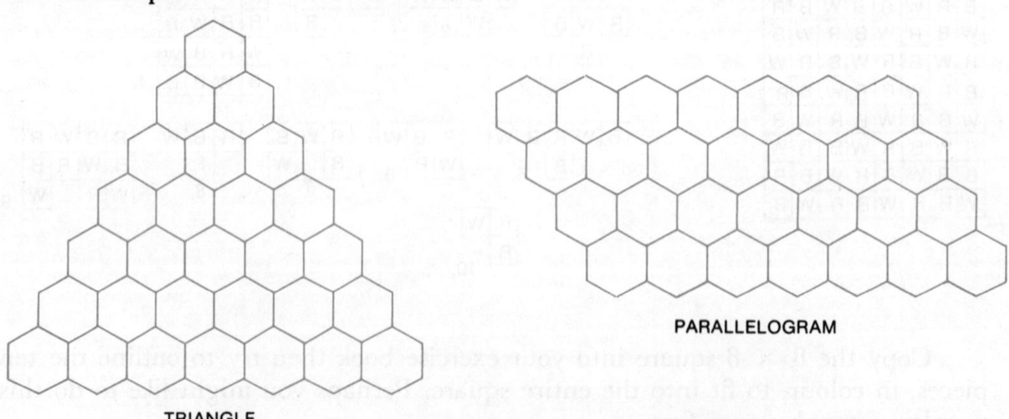

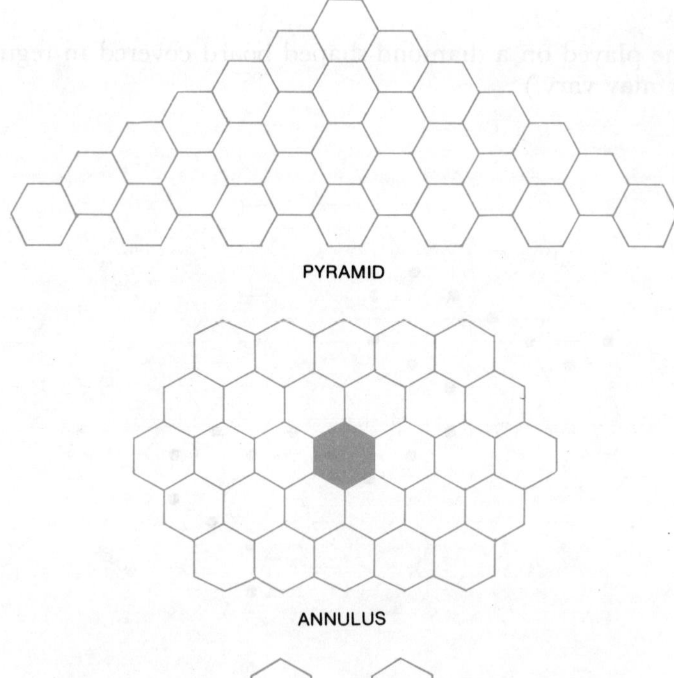

PYRAMID

ANNULUS

BLOT

One of the shapes is impossible.
Can you determine which one?

Pentahexes

The next set has twenty-two different members. Use small isometric grid paper to make a full set.

ACTIVITY

✱ Cut these shapes out of cardboard and fit them together to form a parallelogram or a symmetrical crescent.

There are eighty-two hexahexes (6 hexagons), 333 heptahexes, 1 448 octahexes and 683 101 dodecahexes (calculated by a computer).

It has been said that all polyhexes to order 5 are tilers.

Hex Game

This is a game played on a diamond-shaped board covered in regular hexagons. (The number may vary.)

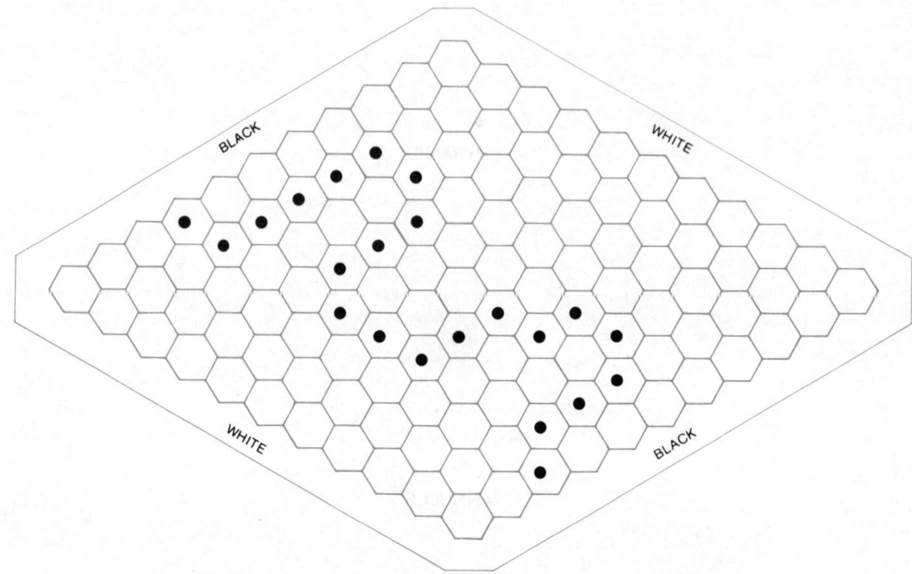

One pair of opposite sides is called "black" and the other pair "white" with the corner hexagons belonging to either colour. Players alternately put a counter on any hexagon which is not occupied. Each player tries to complete an unbroken chain of counters between the two sides. The chain need not be straight (see diagram). Can this game end in stalemate?

Instead of using counters, players may put noughts and crosses on the paper.

Polyiamonds

Following the same principle, polyiamonds are formed by joining equilateral triangles together. The triangular "cousin" to polyominoes, polyiamonds have only been exploited since 1965. The name polyiamond was coined by the Glasgow mathematician, T. H. O'Beirne.

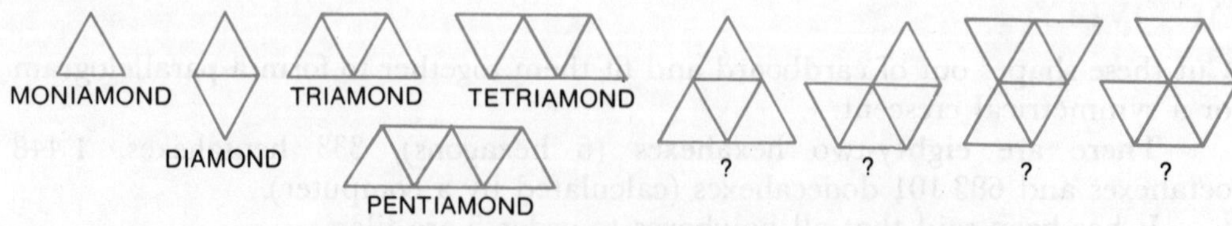

You might like to investigate polyiamonds further. Once again, use isometric graph paper and tabulate your results. Mirror reflections of asymmetrical shapes are not considered different. You should find:

1	diamond	12	hexiamonds (note similarity with pentominoes)
1	triamond	24	heptiamonds
3	tetriamonds	66	octiamonds
4	pentiamonds	160	noniamonds / enneiamonds

Beyond this no accurate counts have been proved.

The Twelve Hexiamonds

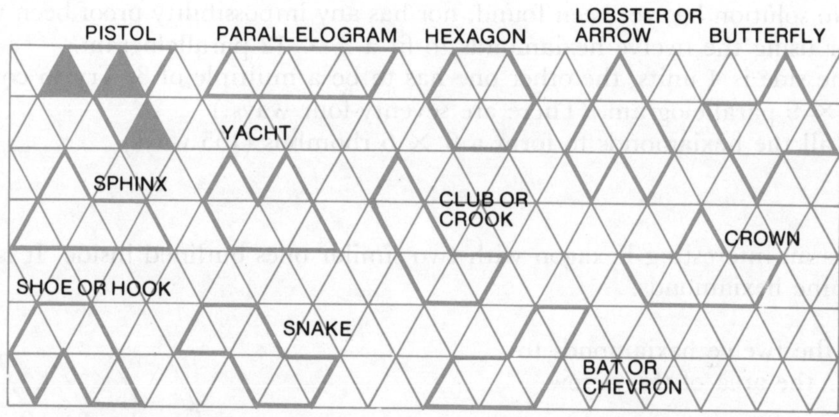

- Which of these shapes are symmetrical? Make a full set on isometric paper then colour alternate triangles (see "pistol" above).
- Which ones have three triangles of each colour?
- Which ones have a different ratio?
- Write down the ratio of the total number of dark triangles to the total number of light triangles.
- Can you explain why it would be impossible to cover this large equilaterial triangle with the twelve hexiamond shapes?

* Use the sphinx, bat, yacht and lobster to form a parallelogram 2 units by 6 units. (This is the only possible parallelogram with one side of 2 units.)
* Now use any four pieces to form a 3 × 4, 3 × 5, 3 × 6, 3 × 7, 3 × 8, 3 × 9 and 3× 10 parallelogram. (A 3 × 3 is impossible.) There are many solutions for these.

A 3 × 11 is possible with eleven pieces. The bat is the shape left out.

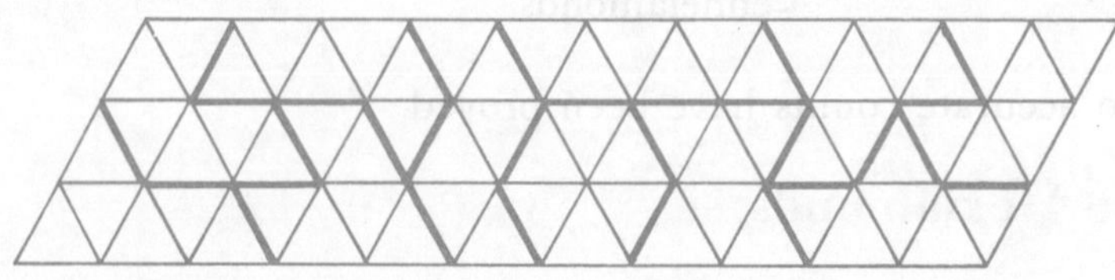

No solution has yet been found, nor has any impossibility proof been worked out for using the twelve hexiamonds to fit a 3 × 12 parallelogram.

* • If one side is 4 units, the other one has to be a multiple of 3. Try to construct a 4 × 9 parallelogram. (There are seventy-four ways!)
* • Use all the hexiamonds to form a 6 × 6 rhombus (155 ways).

Here is an interesting hexagon with two similar ones outlined inside. It is made with nine hexiamonds.

* • Use the twelve hexiamonds to cover the area of this arrow.

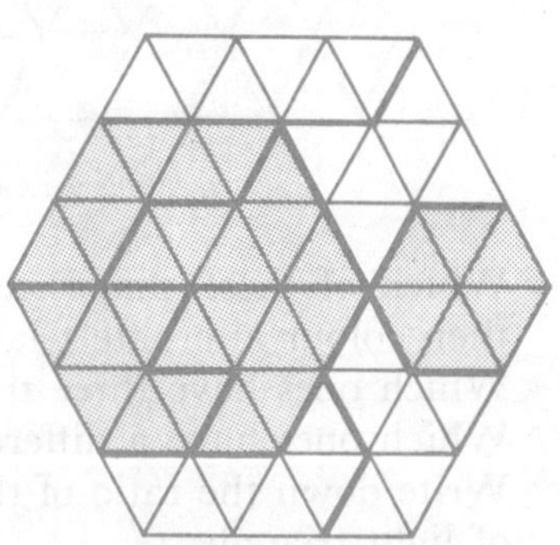

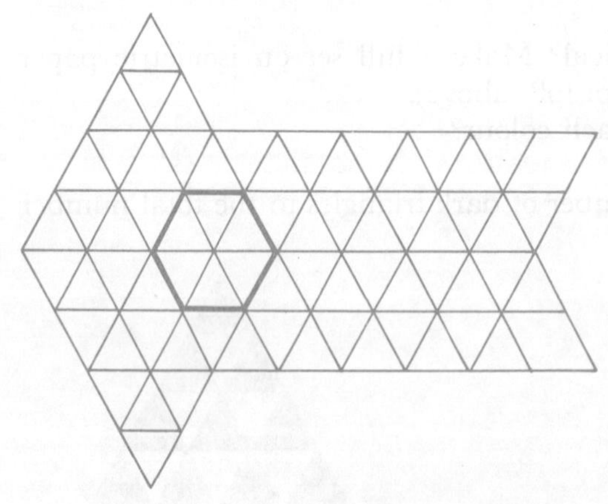

• All polyiamonds of order 6 or less tile a plane, or tessellate. See if you can outline at least six samples of tessellating hexiamonds.

DUPLICATES

Form any shape using two hexiamonds. Make a congruent figure using any other two pieces.

Examples:

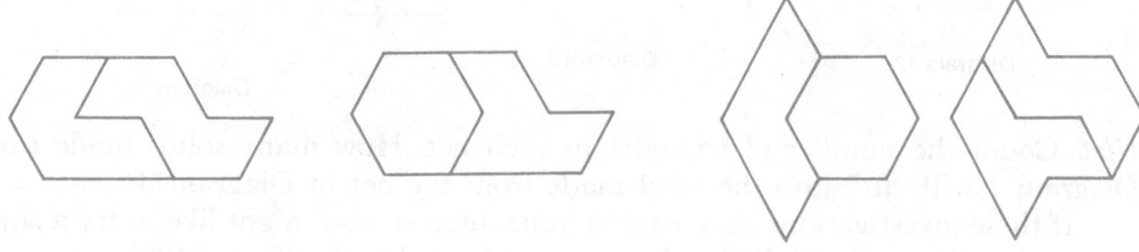

SIMILARITY

Take any shape and use any four pieces to make a similar figure with dimensions twice the original and an area four times it.

Example:

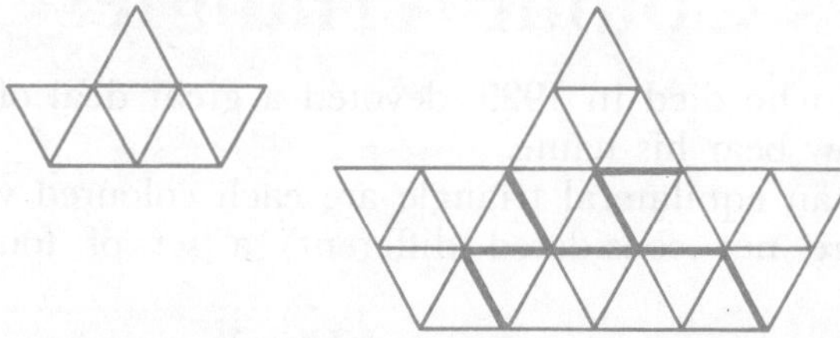

✱ See if you can make the larger version of the butterfly.

- The **Triplication Problem** – forming larger replicas with nine pieces – cannot be solved for the sphinx, butterfly and yacht, but the others are possible. See if you can make a set.

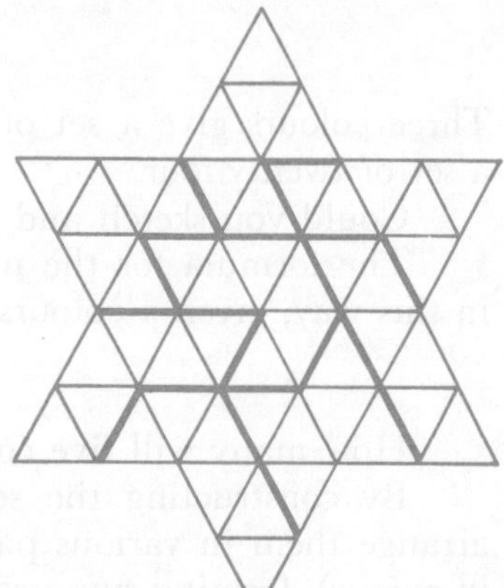

This is a six-pointed star that has the unique 8 piece solution shown.

Could the snake, hexagon or crown ever be on the star's perimeter? Why?

Some Platonic solids (originating from Plato's Academy, around 4th Century B.C.) have nets made of special polyiamonds.
Draw and name the solids shown by these nets:

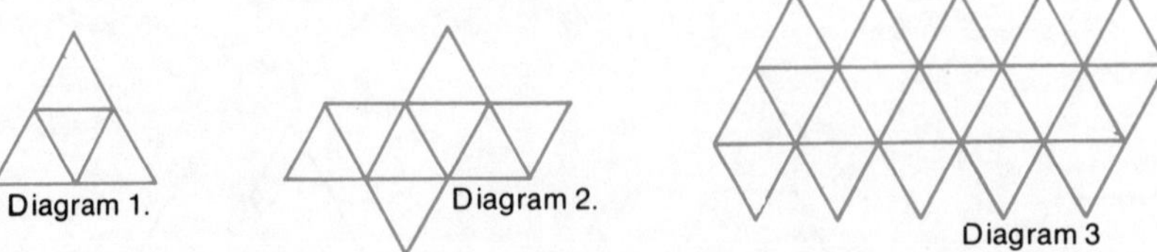

Hint: Count the number of triangles in each net. How many solids made from Diagram 1 will "fit" into the solid made from the net in Diagram 3?

If these investigations have caught your interest, you might like to try a similar one based on a "unit" of a rhombus, with angles of 60° and 120°.

MacMahon's Colour Triangles

Major Percy MacMahon, who died in 1929, devoted a great deal of his life to colour triangles, which now bear his name.

If the three edges of an equilateral triangle are each coloured with one of two colours (rotations are not considered different) a set of four triangles results.

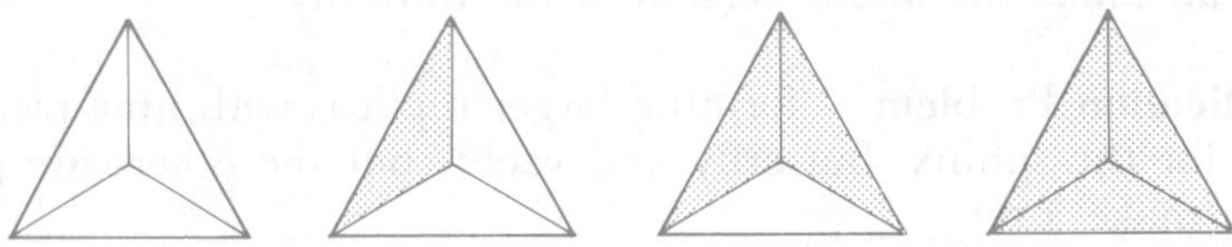

Three colours give a set of eleven distinct triangles while four colours produce a set of twenty-four.

Could you sketch and colour each of these sets?

The formula for the number of equilateral triangles that can be produced in this way, given n colours is

$$\frac{n^3 + 2n}{3}$$

How many will five colours produce?

By constructing the set of twenty-four 4-colour triangles it is possible to arrange them in various patterns, like dominoes (with adjacent edges matching in colour), forming symmetrical shapes. One restriction could be that the entire border of any polygon must be the same colour. Illustrated is a regular hexagon with a perimeter of twelve units, with the border shaded in. Twelve is the minimum-length perimeter of any polygon, and sixteen is the maximum.

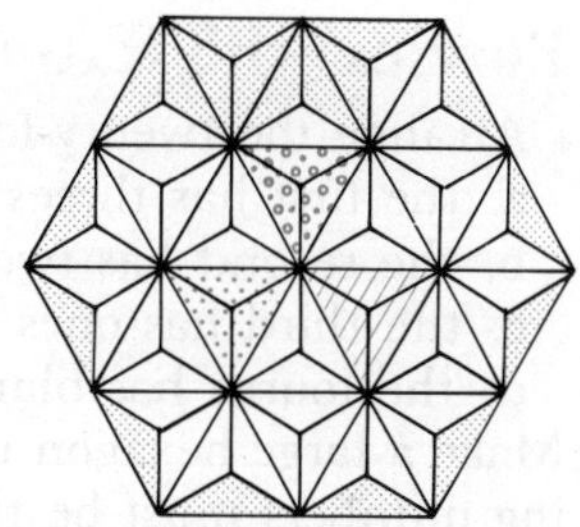

See if you can arrange the triangles to complete this illustration.

It is possible to form a 3 × 4 parallelogram with a perimeter of fourteen units, as well. Try it.

The Game of Trominoes (Trifits)

Dominoes are rectangles divided into two sections. If congruent equilateral triangles, divided into three sections, are used, trominoes are the result.

A set of trominoes can be made from cardboard by dividing congruent equilateral triangles into three sections, then numbering or dotting them from

000 to 333 or 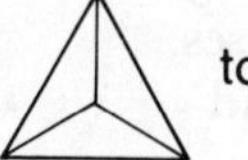 to 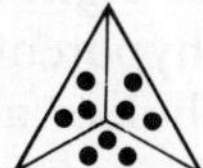

What is the quickest way of constructing six equilateral triangles accurately? (*Hint:* a compass is needed.)

Write down all the combinations in your exercise book before you start putting them on the cards. Be careful not to repeat any. Rotation is allowed, which means that

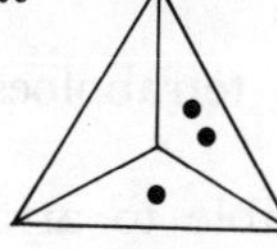 is the same as 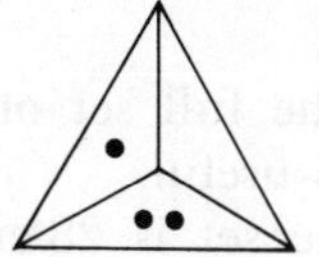 but

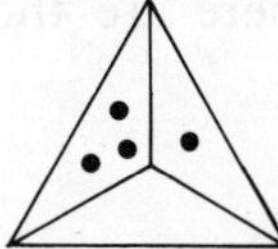 is *not* the same as 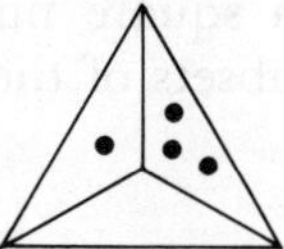

A full set has twenty-four trominoes.

Play the game like dominoes with equal numbers touching one another.

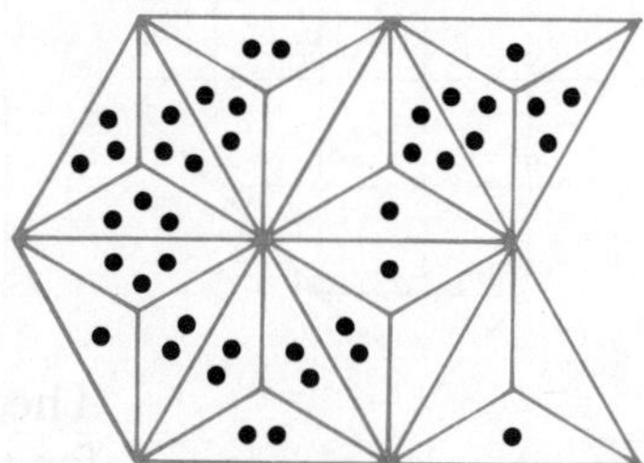

Invent a game for four players with different rules.

Puzzles with Trominoes

- Arrange the twenty-four trominoes into four hexagons so that:
  a. the first has threes all round the edge,
  b. the second has twos all round the edge,
  c. the third has ones all round the edge,
  d. the fourth has blanks all round the edge.

* Make a large hexagon using all twenty-four trominoes. The condition that touching numbers must be the same, still applies.

  Can you now rearrange them so that the sum of two touching numbers is three?

- Make an equilateral triangle using sixteen pieces.
- Make a parallelogram with eighteen pieces.
- Make a trapezium using twenty-one pieces.

Polyaboloes

One of the more recent investigations has been into polyaboloes, irregular polygonal units using such figures as isosceles right-angled triangles. They can be joined either at their sides or along their hypotenuses.

There are only three diaboloes, four triaboloes and fourteen tetraboloes.

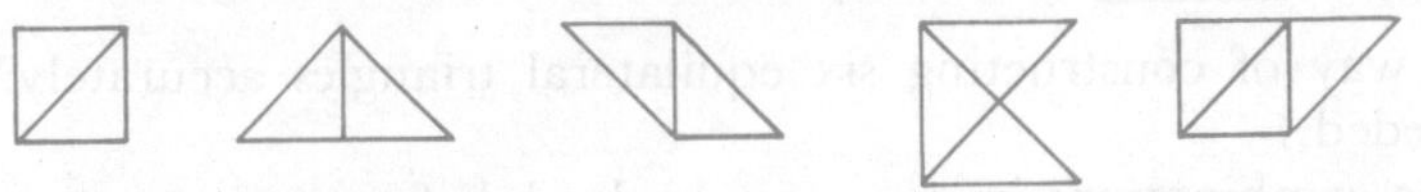

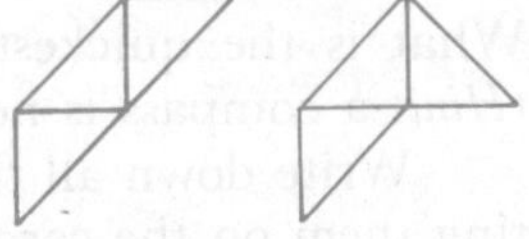

ACTIVITIES

* In your book, work out the full set of fourteen distinct tetraboloes. For this exercise 1 cm grid paper is useful.

  As the total area of the set is 28 cm^2, it is not possible to arrange them into a square. (28 is not a square number.) However, there are three squares that can be formed with subsets of the complete set:

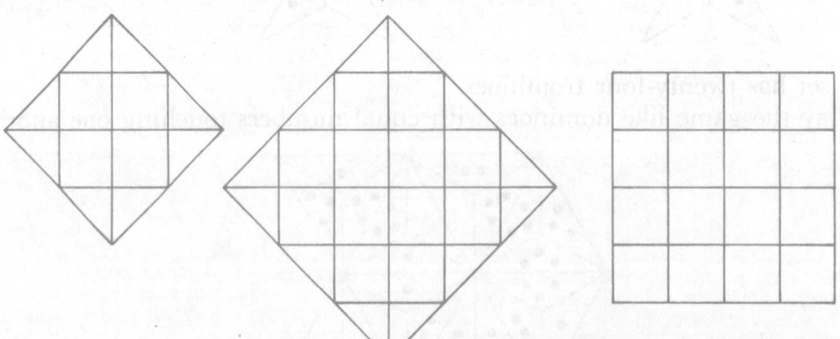

There are two solutions for this square.

The number of solutions for these squares has not yet been determined.

* Use three pieces to form a 2×3 rectangle.
* Make a 2×4 rectangle using four tetraboloes.

Experiment with other subsets to form more rectangles with different dimensions. Try two distinct types: those with perimeters formed by sides only; those with perimeters made from hypotenuses only.

Examples:

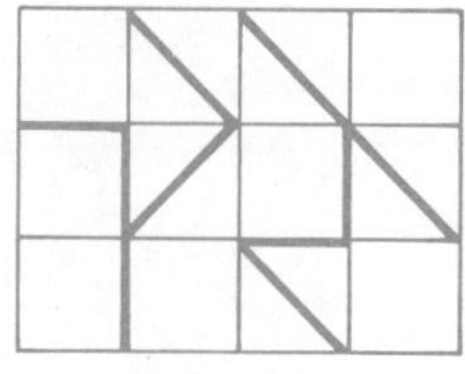

Rectangle with a border made of sides only.

Rectangle with a border made of hypotenuses only.

As the topic of polyaboloes is so new, insufficient work has been done on pentaboloes to be certain of the exact number, but it has been expressed as 30, with 107 hexaboloes.

Polyominoids

Allied to polyominoes, polyhexes and polyiamonds (all with only two dimensions, length and breadth) there are three dimensional shapes called polyominoids.

Instead of using unit squares, as in polyominoes, cubes can be joined together face to face. Centicubes are ideal for building up a set of polyominoids.

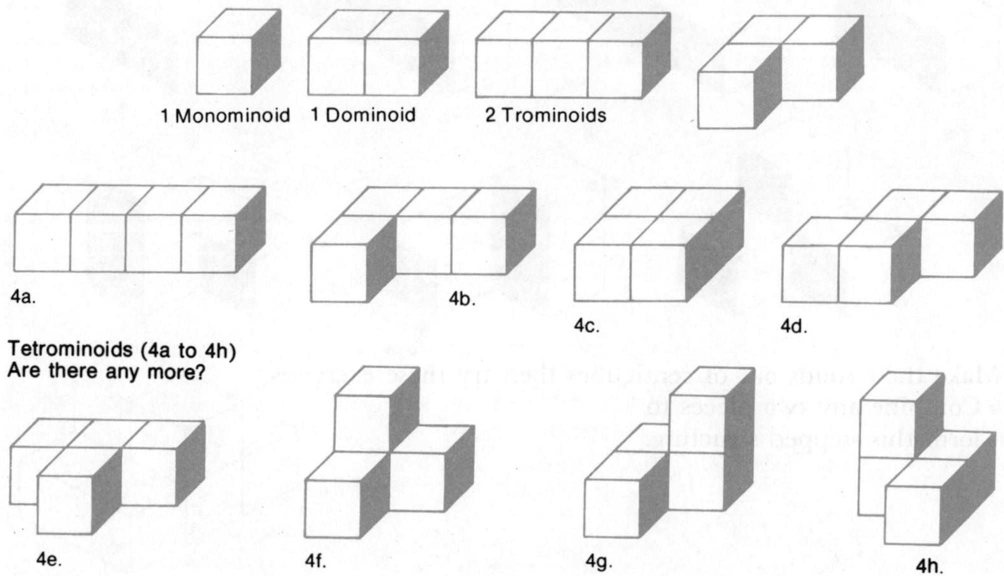

- Is it possible to form the tetrominoids into a rectangular block? You may form a 4 × 4 × 2 or 2 × 2 × 8 block.
- Try constructing a symmetrical structure, such as a set of steps, with all the tetrominoids.
- How many pentominoids can you make? You may need a lot of patience to make 29 or 166 hexominoids.

Since there is no way to "turn over" an asymmetrical polyominoid, mirror-image pairs are considered to be different.

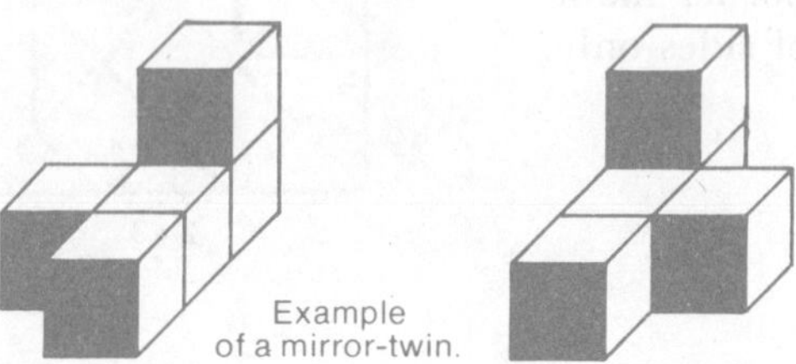
Example of a mirror-twin.

The Soma Cube

Soma: Greek meaning the body. Biologically, the axial part of the body, without limbs.

This is a 3 × 3 × 3 cube dissected into six pieces composed of 4 unit cubes and one piece of 3 unit cubes, invented by a Danish writer and puzzle expert, Piet Hein.

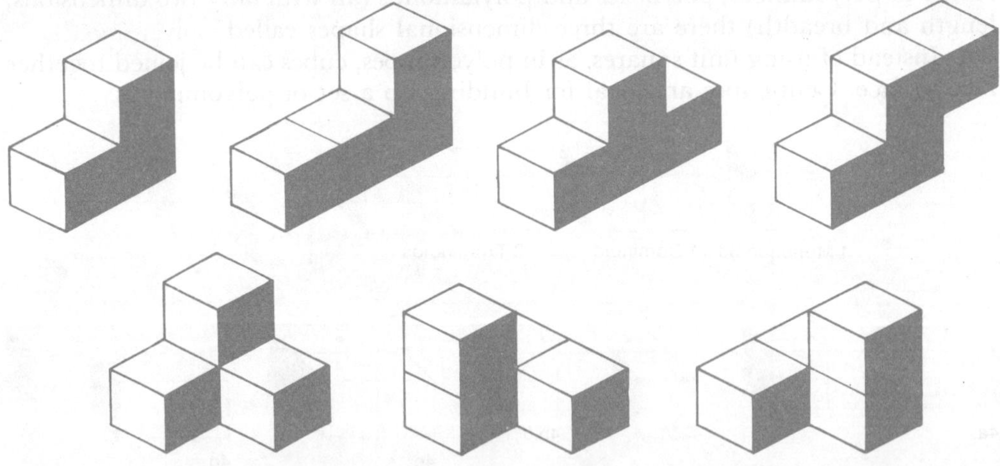

Make these solids out of centicubes then try these exercises:

- Combine any two pieces to form this stepped structure.

• Now use all seven pieces to form the following couples.

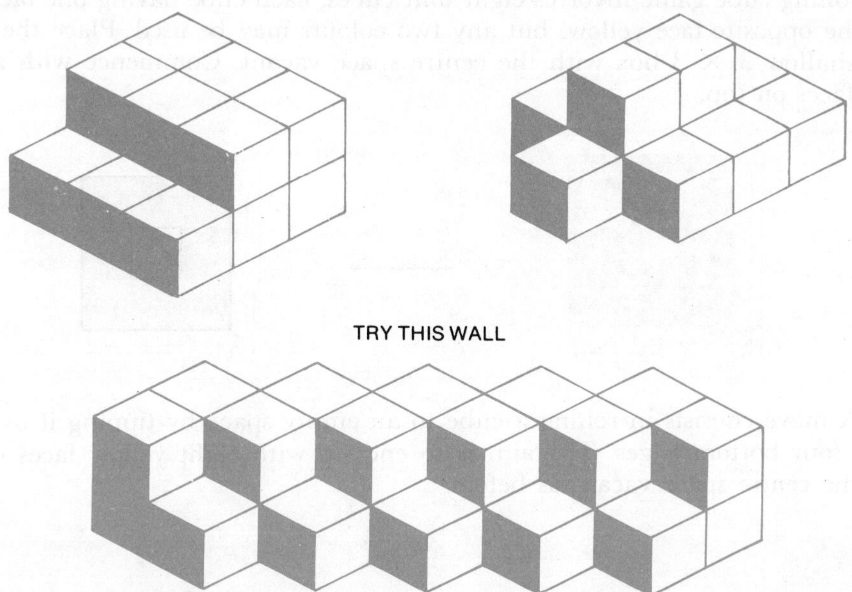

TRY THIS WALL

• Now try to make the 3 × 3 × 3 cube. Set the more irregular shapes first and leave the simplest one till last.

If you are successful, make some other intriguing shapes of your own.

The Soma puzzle is not the first polycube dissection of the 3 × 3 × 3 cube to be marketed as a puzzle. A six-piece one was sold in Victorian England under the name of the **Diabolic Cube.** All the pieces were planar and contained one of each of the orders 2 to 7.

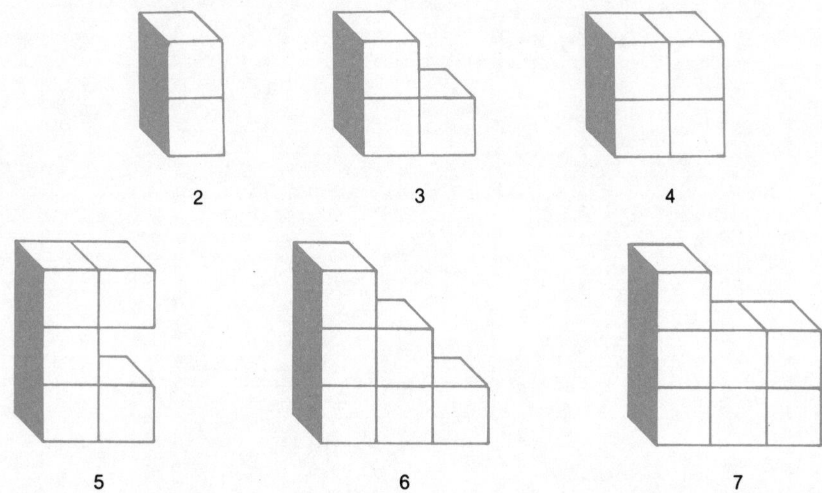

There are thirteen ways the Diabolic Cube can be assembled.

For Patient People!

The rolling cube game involves eight unit cubes, each cube having one face blue and the opposite face yellow, but any two colours may be used. Place the cubes in a shallow 3 × 3 box with the centre space vacant. Commence with all the blue faces on top.

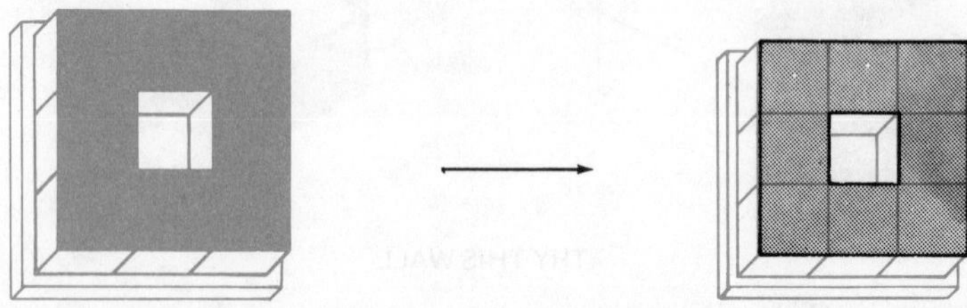

A move consists in rolling a cube to an empty space by tipping it over one of its four bottom edges. The aim is to end up with eight yellow faces on top and the centre space vacant as before.

Unit 7 Tessellations-Mathematical Mosaics

Imagine you are given an infinite supply of jig-saw puzzle pieces, all congruent. If it is possible to fit them together without gaps or overlaps, by reflecting, rotating or translating, to cover an entire plane the result is said to be tessellating. From ancient times such tessellations have been used for floor and wall patterns as well as on rugs, tapestries, quilts, clothing, furniture and for wire netting. In industry, where would tessellations be used?

Pappus, who lived in 4th Century B.C. recognised that bees use the regular hexagon exclusively for the shape of their cells in the honeycomb. This is the strongest and most economical shape of all.

The Greeks proved that only three regular polygons will tessellate: the equilateral triangle, the square and the hexagon. However, it was probably Kepler (1571–1630) who further investigated the many possible ways of filling a plane with regular polygons.

The simplest regular polygon is the equilateral triangle.

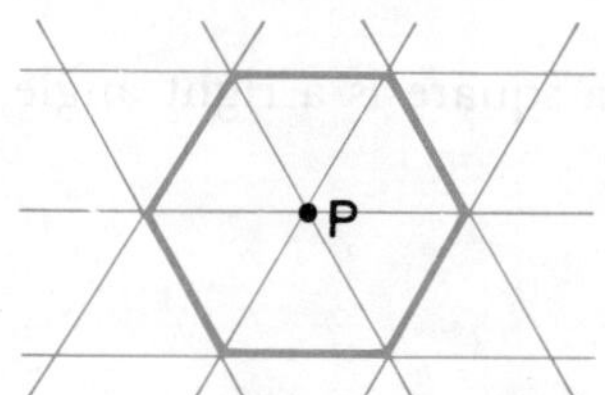

Here, six triangles fit exactly around a point P. What must be the size of each angle around this point?

An infinitely varied set of patterns can be formed by "sliding" rows of triangles along the lattice lines, to produce designs like these:

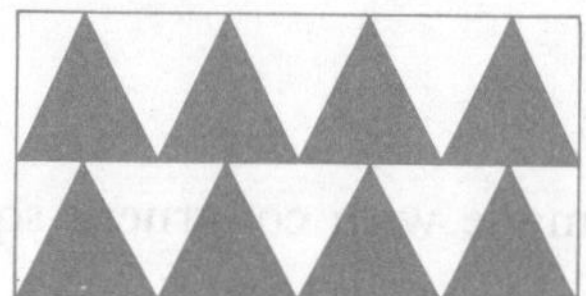

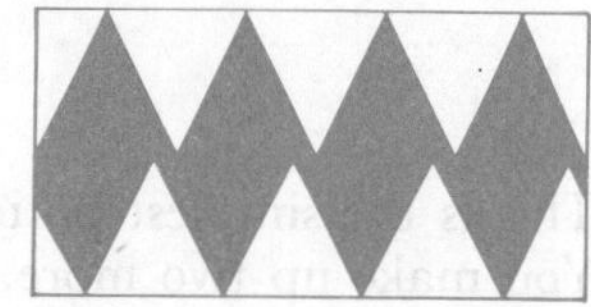

This pattern of equilateral triangles was used for the floor of a Roman house in the 1st Century, A.D.

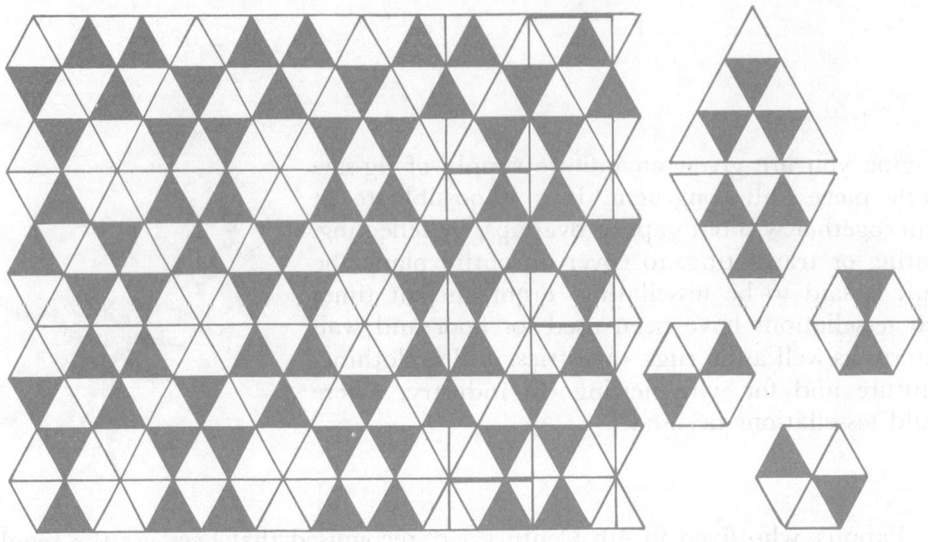

It is interesting to break the pattern down into repeats. Can you find:

- parallelograms, containing twelve light tiles and six dark tiles?
- triangles, composed of six light tiles and three dark tiles (arranged in two different directions)?
- hexagons, containing four light and two dark tiles (arranged in three different directions)?

Can you also find other triangles, quadrilaterals and hexagons?

Another possibility is the square. As each angle of a square is a right angle, four will fit exactly around a point.

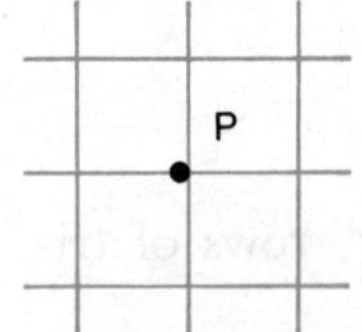

This familiar pattern of squares is found on a chess board, or on a sheet of grid paper.

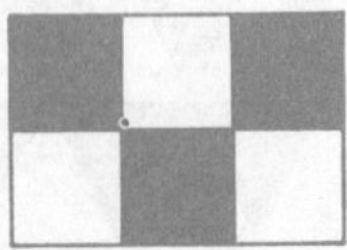

This is the simplest pattern made with congruent squares. You make up two more.

Designs with hexagons are fixed (think of a piece of honeycomb) because there is only one set way to tessellate them. Of the three regular polygons, the hexagon covers the greatest area for any particular given perimeter. Each vertex of a hexagon is surrounded by exactly three hexagons.

How many degrees will the interior angle of each hexagon be?

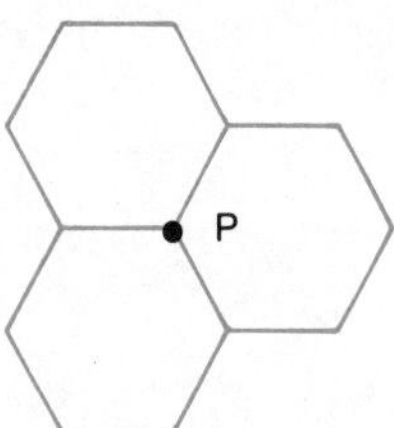

If other regular shapes (with all sides equal) are used, there are some interesting arrangements formed. Take an octagon for example.

- Using the formula $(2n - 4)$ right angles, calculate the sum of the eight interior angles, where $n = 8$.
- Divide this total by 8 to work out the size of one interior angle.
- If two octagons are joined edge to edge (as in the diagram) can you then work out the size of the remaining angle?

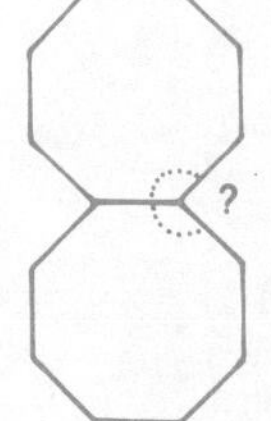

Only two octagons can meet at a point, with a small space "unfilled".

- What geometrical shape will therefore "fit" in between regular octagons?

Draw the repeating pattern into your book. It is called a complex mosaic, because it involves two shapes.

There are eight such semi-regular tessellations made up of different combinations of triangles, squares, hexagons, octagons and dodecagons. Each vertex has two or more kinds of regular polygons fitted together, corner to corner, in a cyclic order.

It is convenient to use the Schläfli symbol $\{p, q\}$ where p stands for the number of vertices and q indicates the number of shapes surrounding one vertex, for tessellations of regular polygons.

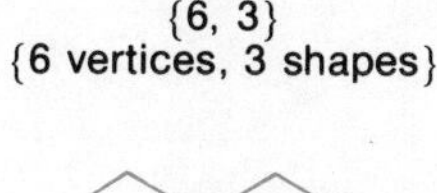

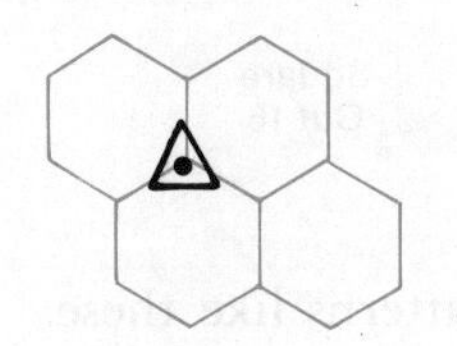

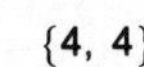

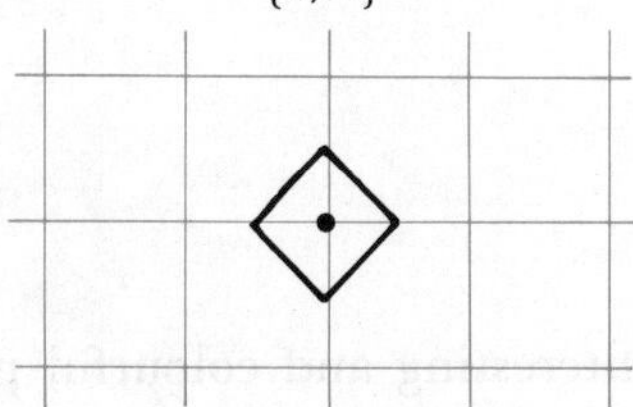

{3, 6}

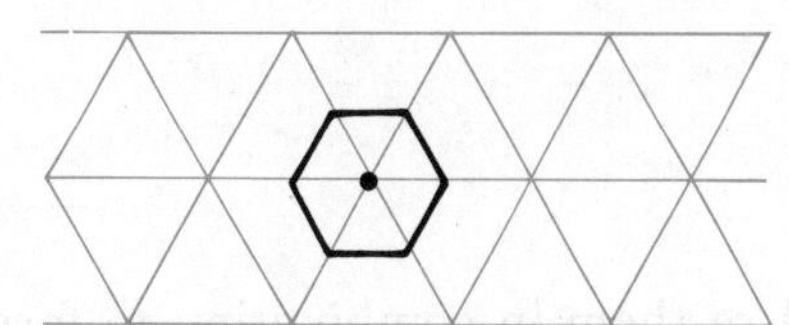

If the sides of each hexagon are perpendicularly bisected, a dual is formed, with the symbol {3, 6}.

What is the dual of {4, 4}?

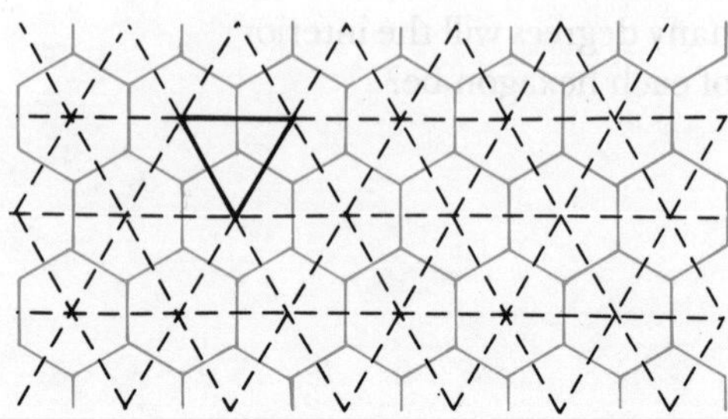

ACTIVITY

Draw accurately, or trace, these five regular polygons, then cut them out as templates and use them to produce the given number of each. Use coloured paper.

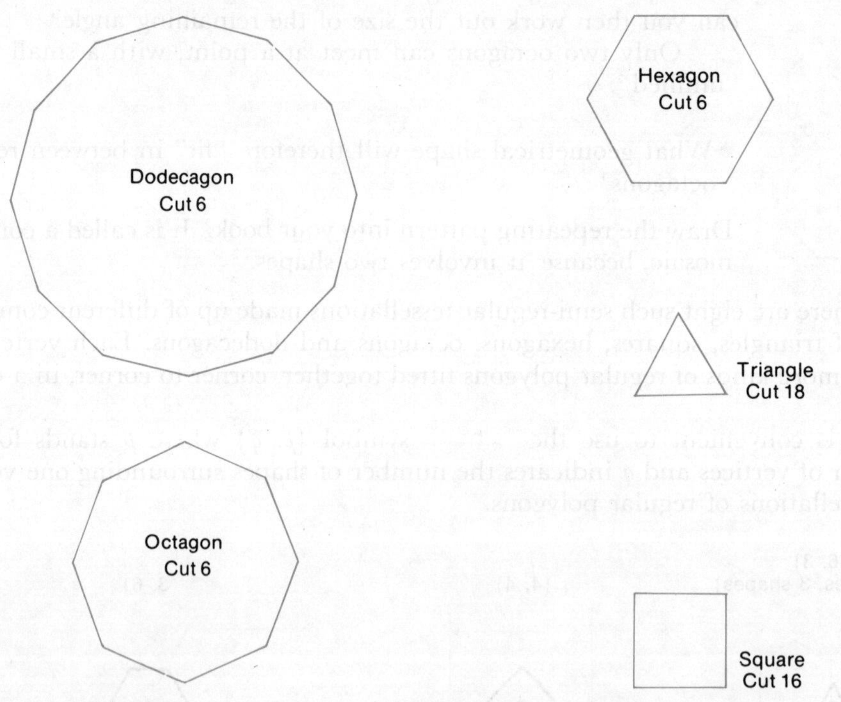

Use them in combination to form interesting and colourful patterns like these.

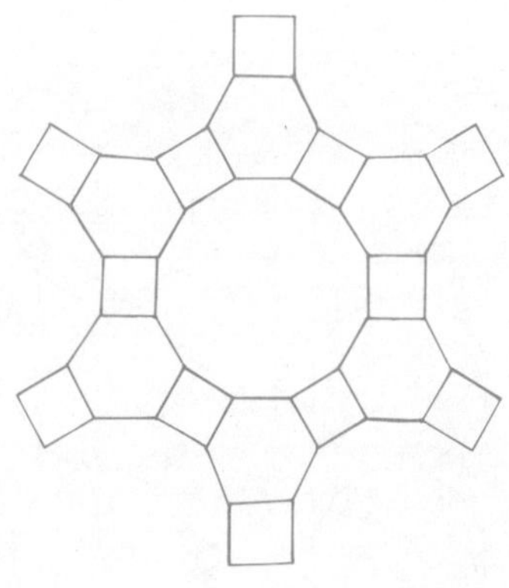

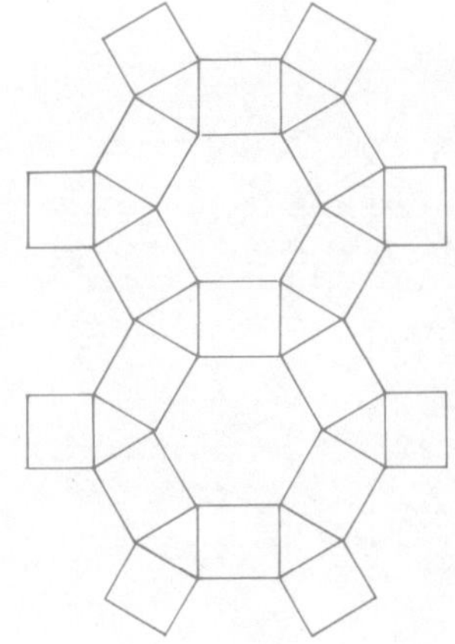

Do you recognise
this one?

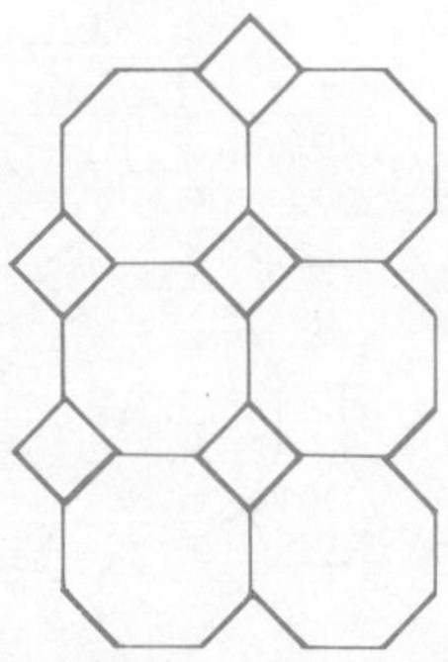

The Eight Semi-Regular Tessellations

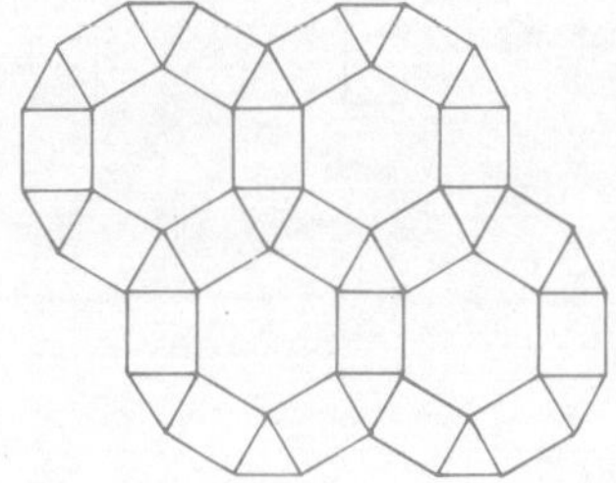

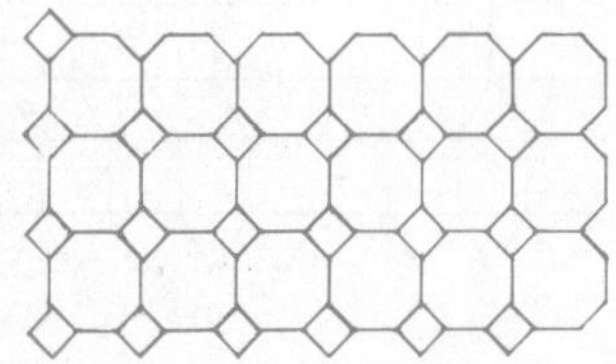

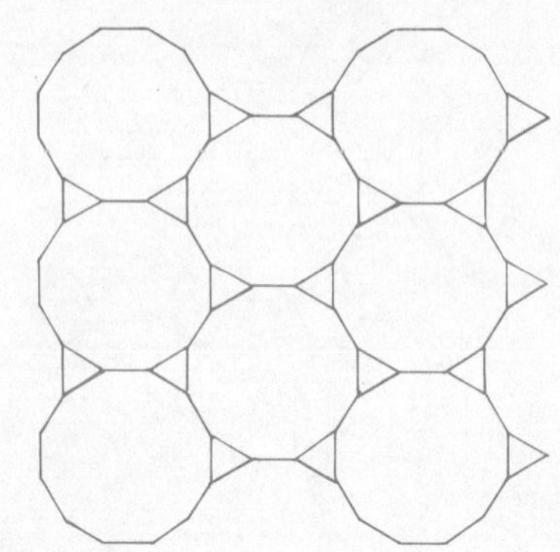
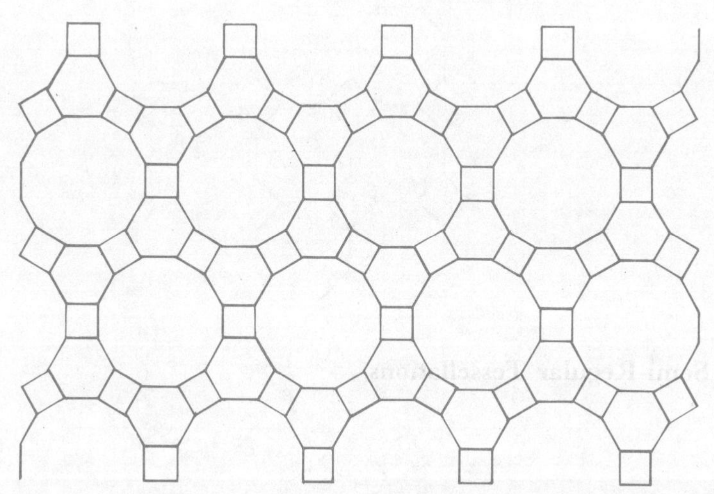
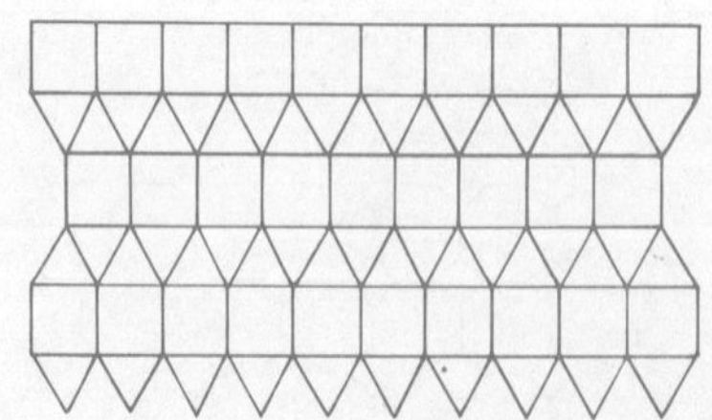

If a regular polygon is going to be used as a tessellation an integral number of vertices must fit together at a point; that is, the size of each interior angle must be a factor of 360°. Those over 60° are 72°, 90°, 120°, 180° and 360°.

The size of the angle is a function of the number of sides and can be calculated from the formula

$$a = 180 - \frac{360}{n}$$

where a = angle, in degrees and n = number of sides.

Substitute $n = 3$ (triangle's sides) to calculate the size of the angle at a point. Do the same when $n = 4, 5, 6, 8$ and 12.

You should find that as the number of sides increases, so does the size of the angle.

Although only three regular polygons can be used to form tessellations containing only one shape, many irregular shapes will also tessellate.

Use grid paper where applicable to copy and continue some of these common geometrical ones. Include any variations of your own.

Irregular Tessellations

Type 1.

a. Congruent triangles of any shape can be used for a tessellation.

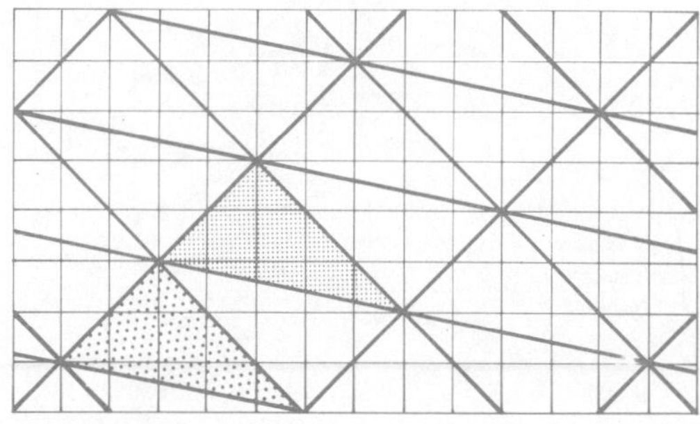

b. Congruent quadrilaterals will tessellate.
Count the squares carefully!

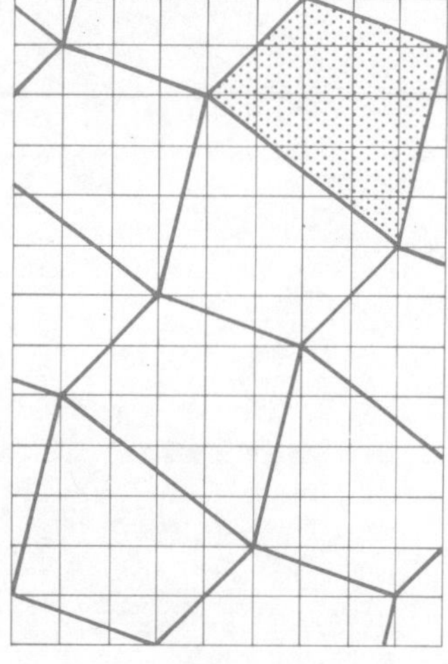

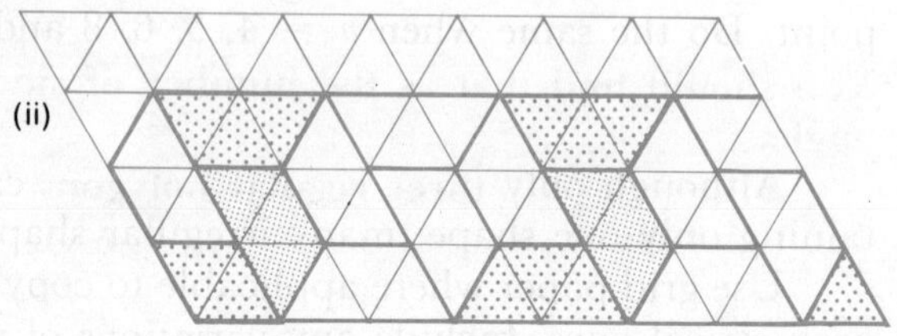

Take a pair of identical quadrilaterals (one inverted with respect to the other) then join the corresponding edges to create a hexagon. This hexagon, when translated, will form a tessellating pattern.

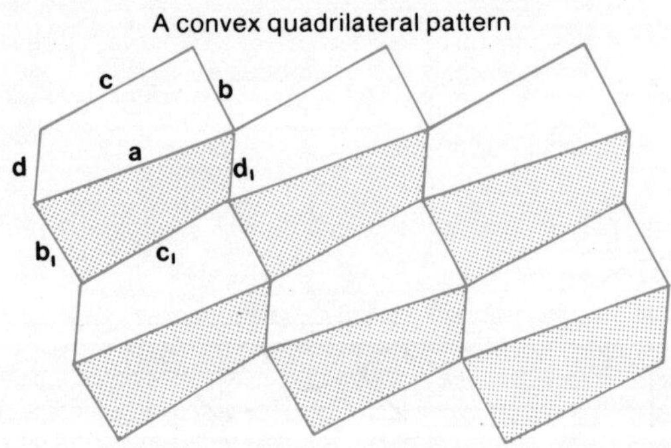

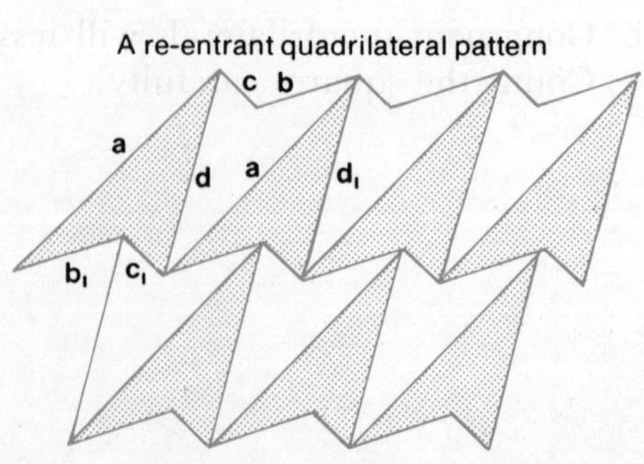

c. Congruent parallelograms are easier.

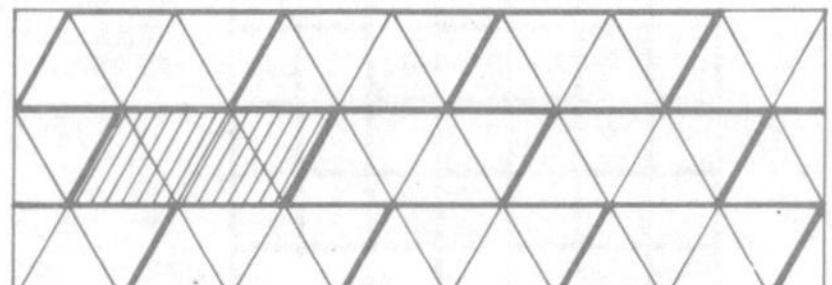

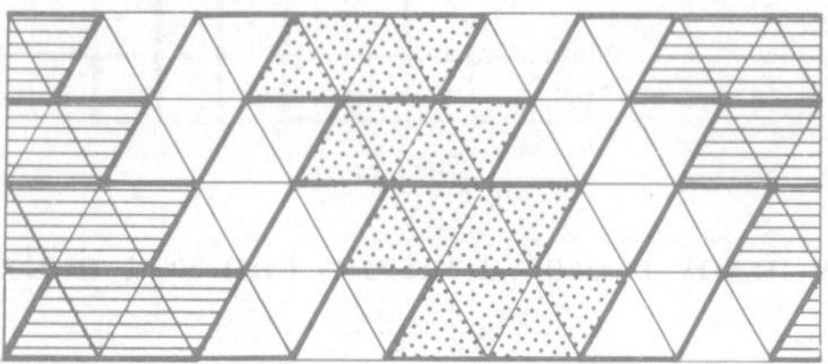

Rectangles are used extensively in houses, in the brickwork and on parquet floors.

The simplest

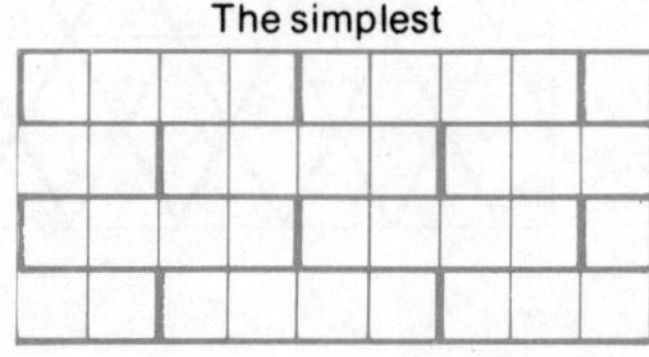

A variation

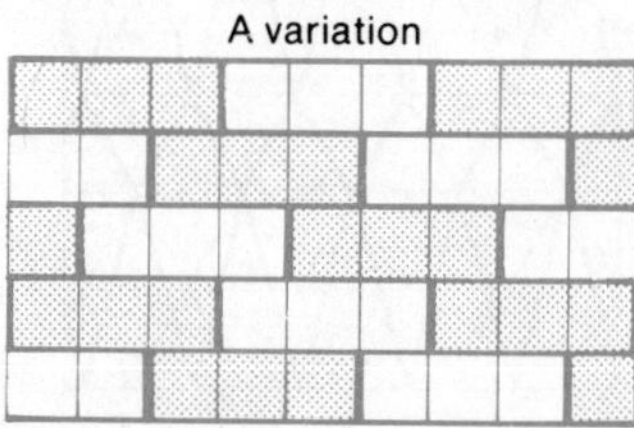

Parquet

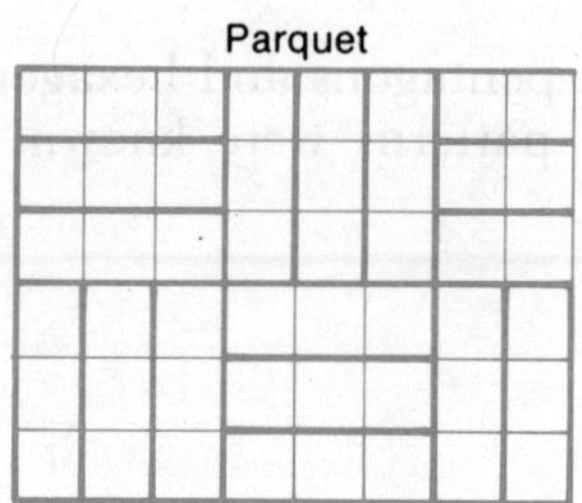

Simple parquet

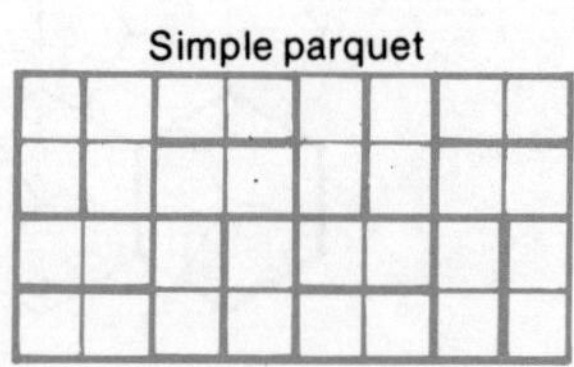

A haphazard arrangement

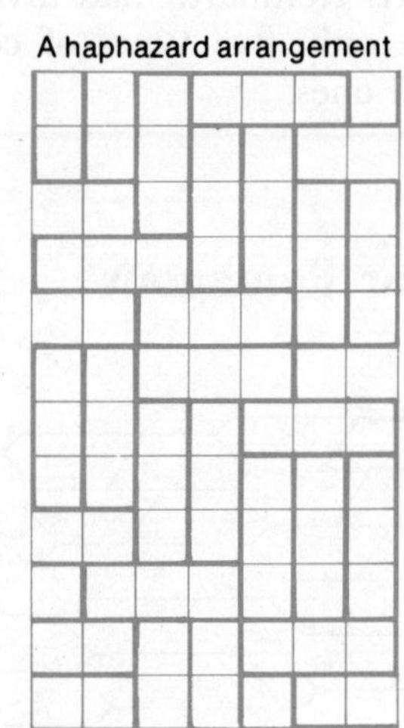

More parquet variations—herringbone pattern

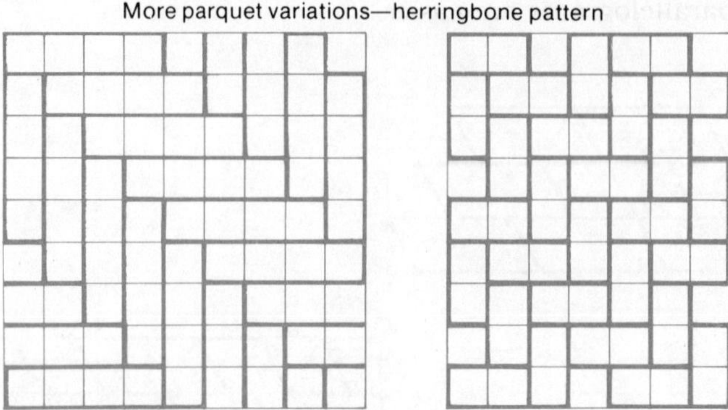

How many more variations can you make?

"Kite" tessellation

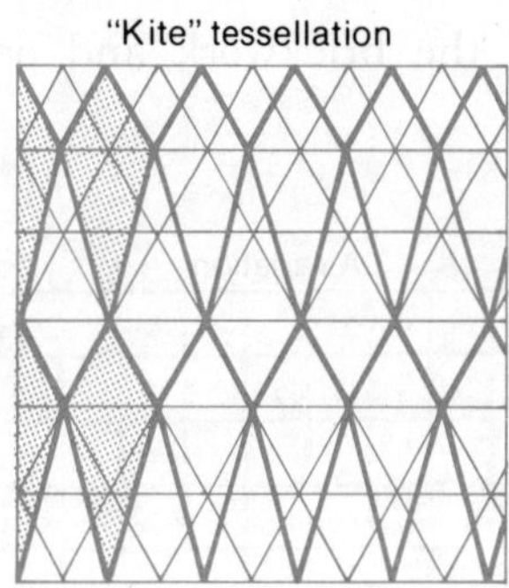

Rhombus tessellation

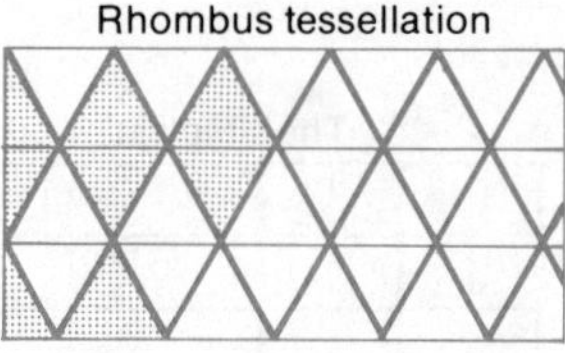

By 1918 K. Reinhardt had investigated convex pentagons and hexagons. Up till that time only five types of convex pentagon patterns were known and three hexagonal ones.

Irregular Pentagons

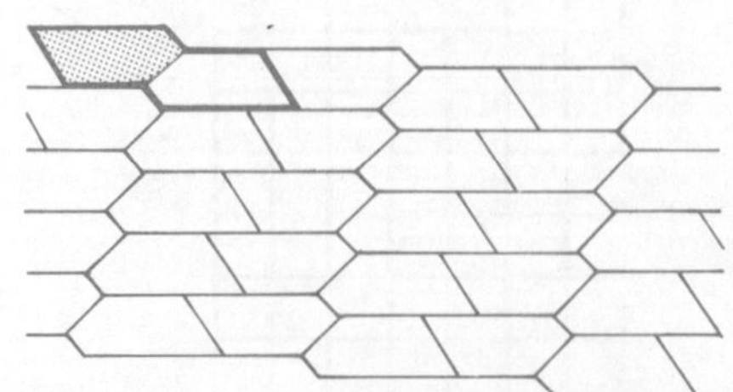

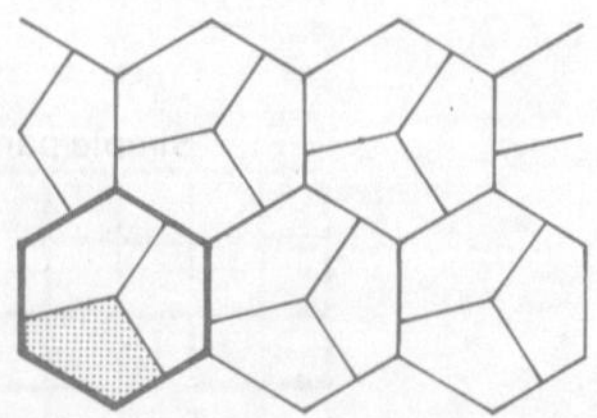

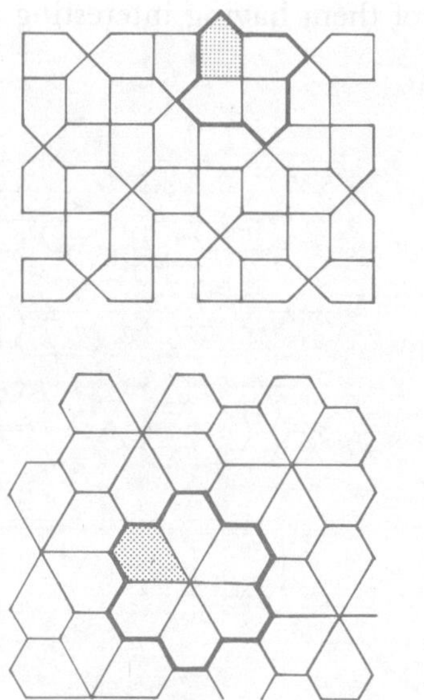

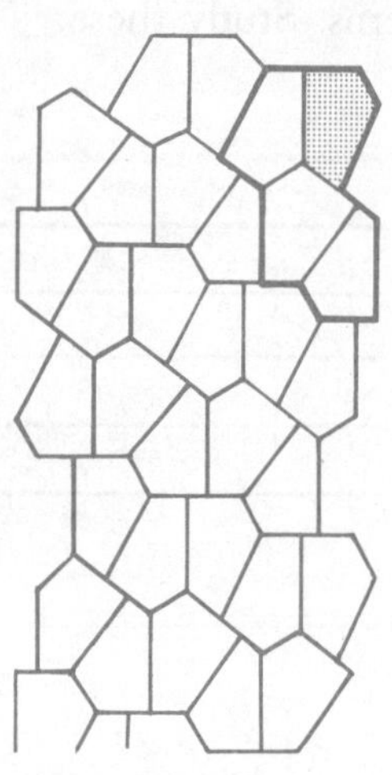

The following three types of convex pentagonal tessellations were discovered in 1967.

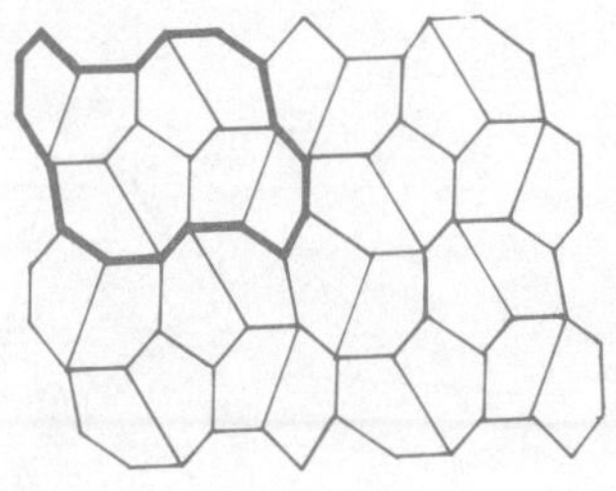

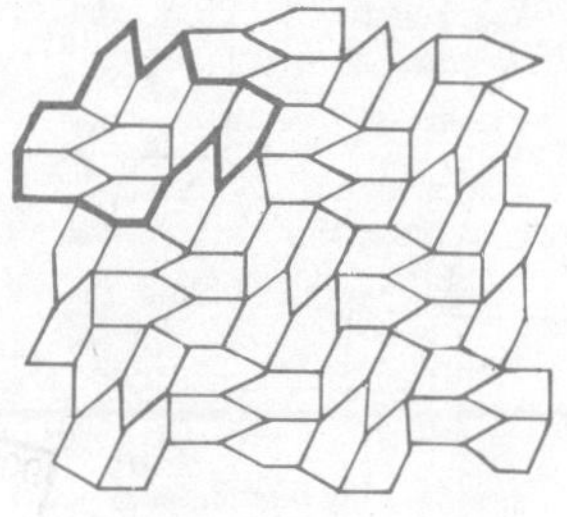

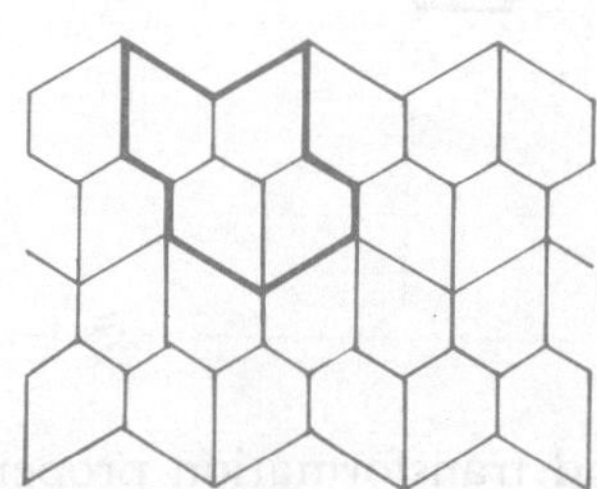

There are many other possibilities with most of them having interesting features and patterns. Study these:

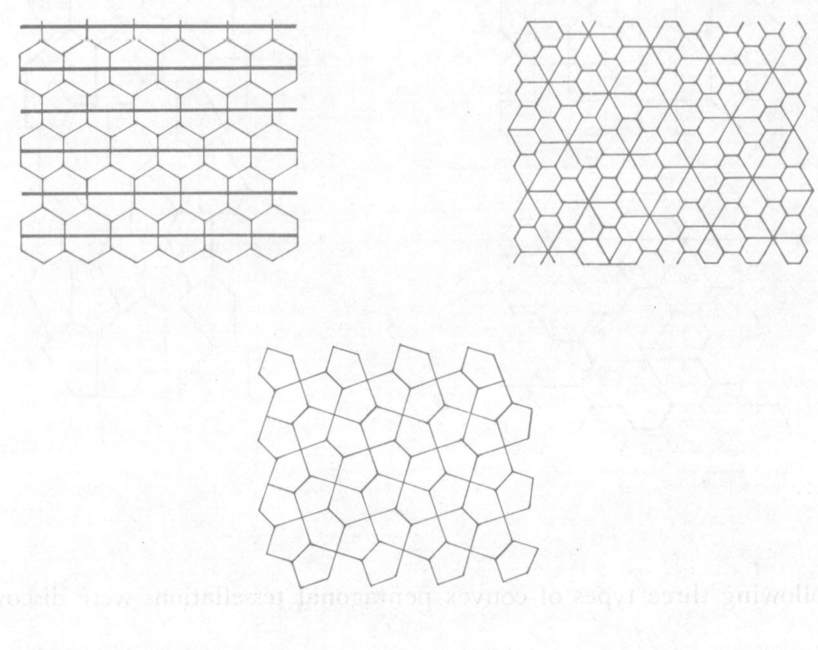

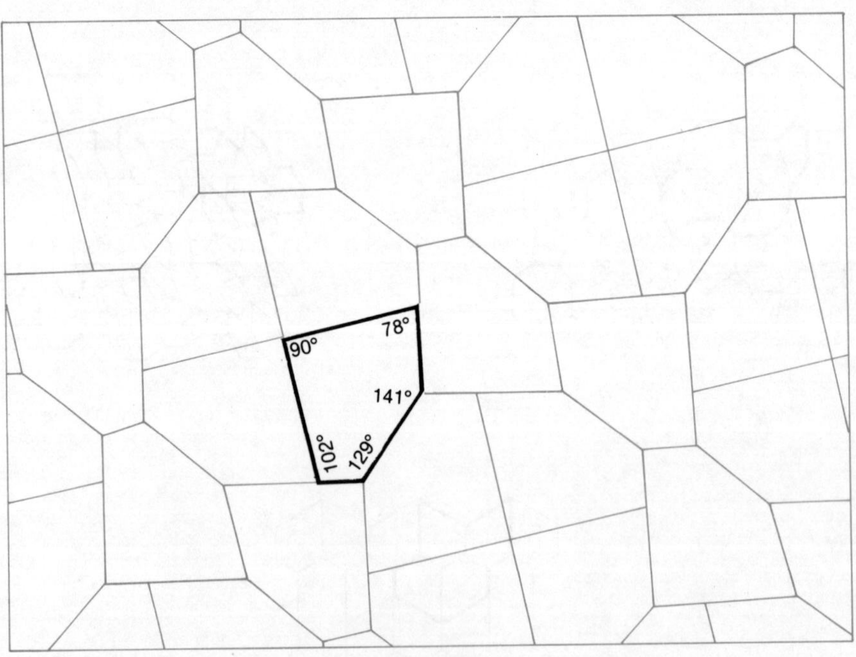

Discuss their symmetrical and transformation properties.

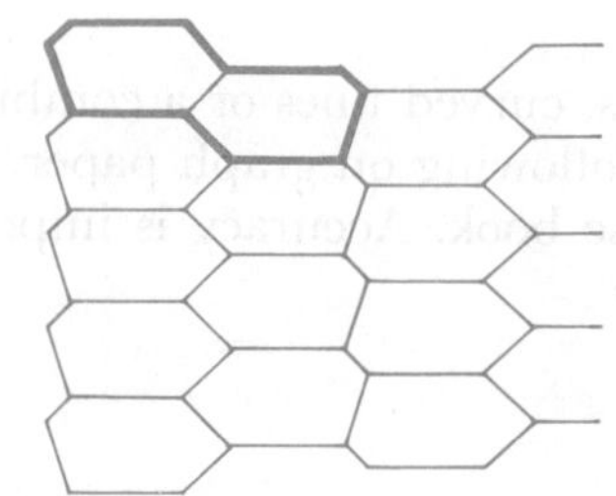

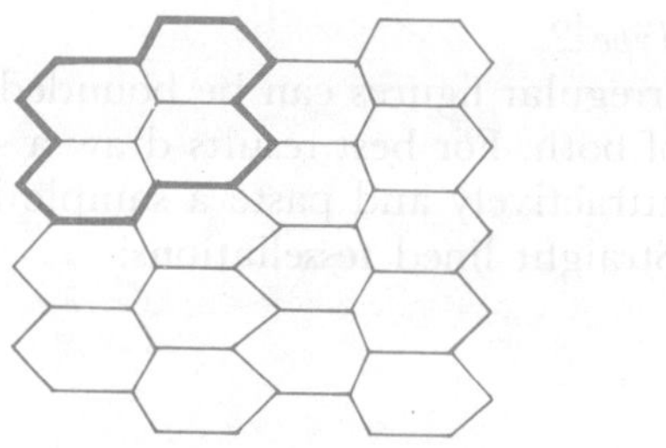

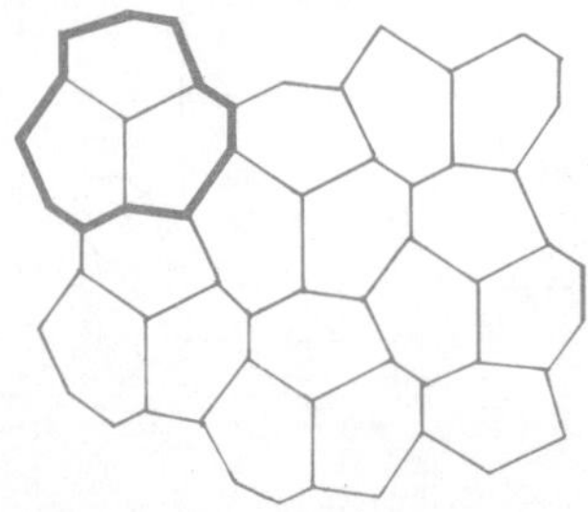

Three types of irregular **hexagon** tiling arrangements.

Hexa-Pentagonal Crossword Puzzle by Croton

Here are sixty-four pentagons, all tessellating, which form "cells" – sixteen horizontal hexagons intersected by nine complete vertical hexagons, with portions of others. Can you see them all?

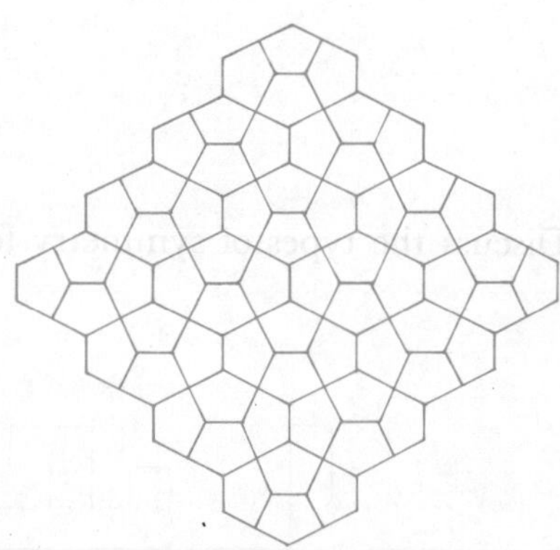

This completed crossword contains many words. Write down all the words that are:

- in each hexagon (4 letters each), reading clockwise.
- formed by the letters in the eight pentagons along each of the four boundaries (corner to corner in a clockwise direction).

Look up any words you don't know.

Type 2

Irregular figures can be bounded by straight lines, curved lines or a combination of both. For best results draw a selection of the following on graph paper, colour attractively and paste a sample into your exercise book. Accuracy is important. Straight lined tessellations:

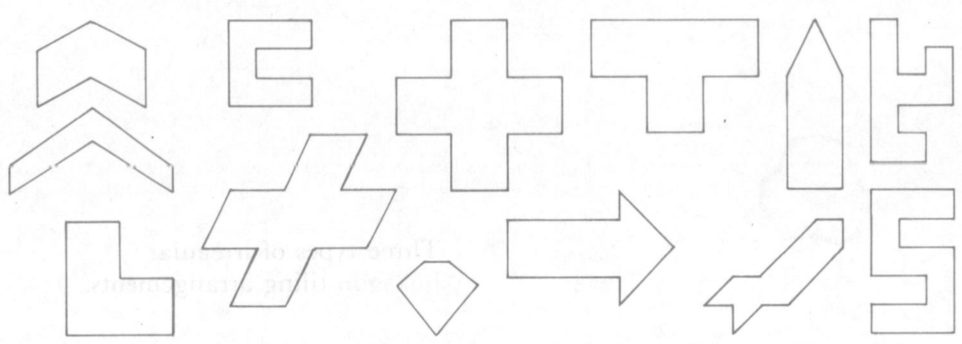

Discuss the types of symmetry found in these shapes.

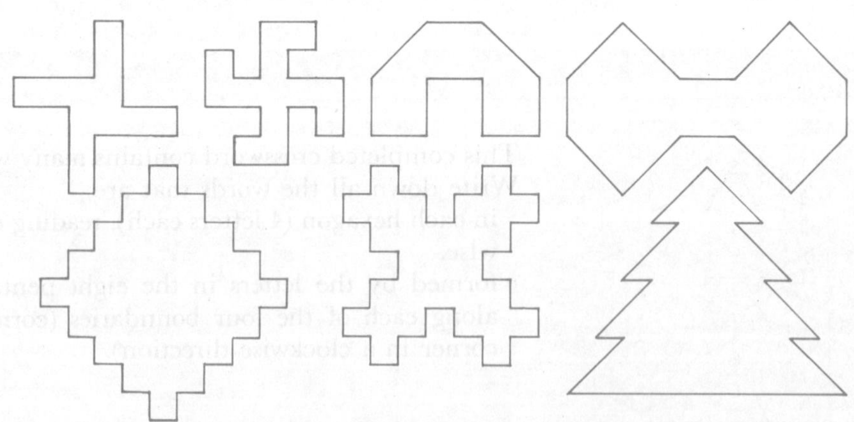

More complex tessellations.

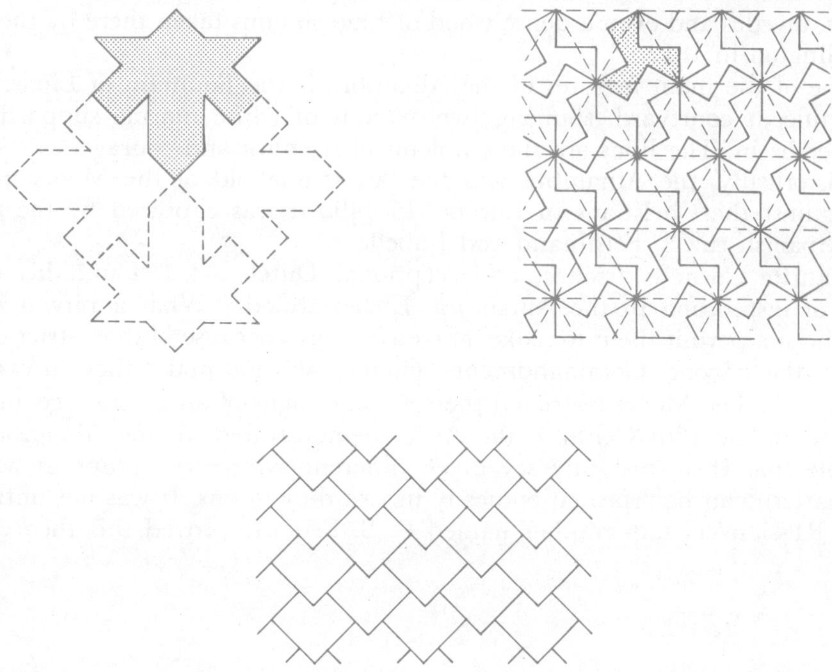

These two tessellations are found in the Alhambra in Spain.

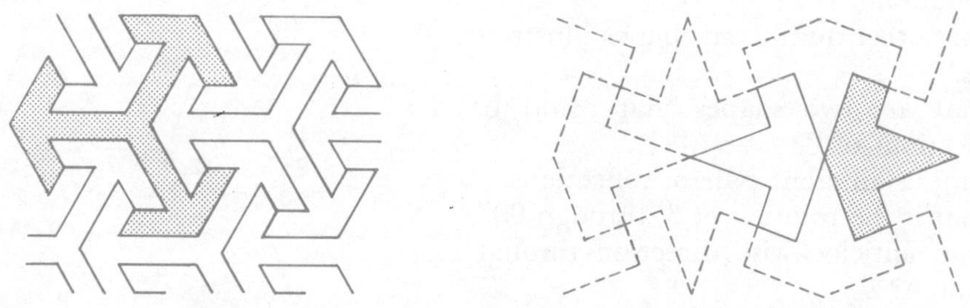

The Alhambra (from an Arabic word literally meaning "The Red")

This is an ancient palace and fortress at Granada, in Spain, that has many brightly-coloured, tile-covered walls, floors, halls and courtyards, with many of the designs being tessellations. The Alhambra was built by the Moors between 1248 and 1354, in the reigns of Al Ahmar and his successors. It is one of the finest examples of Moorish art in Europe. The beautiful interior decorations are ascribed to Yusuf I, who died in 1354.

Enclosing fourteen hectares is an outer wall made from red bricks, which contains thirteen towers. The grounds are exquisite, being planted with roses, oranges, myrtles and even a dense wood of English elms taken there by the Duke of Wellington in 1812.

One of the main features of the Alhambra is the Fountain of Lions, set in a magnificent courtyard, showing twelve lions of white marble supporting an alabaster basin. The lions are the emblems of strength and courage.

Historically, the Alhambra was the last stronghold of the Moors in their fight against the Christians in Europe. In 1492 it was captured by the armies of the Spanish rulers, Ferdinand and Isabella.

Maurits Cornelis Escher, an exceptional Dutch artist, found his richest source of inspiration in the Alhambra. Escher stated: "What a pity it is that Islam did not permit them to make 'graven images' (because of their strict observance of the Second Commandment 'Thou shalt not make thee any graven image . . .')". The Moors restricted their art to designs of an abstract geometrical type and in the 13th Century, the Arabs demonstrated, in the Alhambra, the principle that there are only seventeen different symmetry groups in which a basic pattern can be repeated endlessly in two dimensions. It was not until 1891 that a Russian crystallographer named E. S. Fedorov proved this theory to be true.

TRANSFORMATION EXERCISE

- 1 and 2 illustrate what type of transformation?
- What other numbered shapes illustrate this?
- What are two shapes that could be reflections of 8?
- Name a different pair of reflections.
- What is the rotation of 3, through 90°, in an anticlockwise direction through point A?

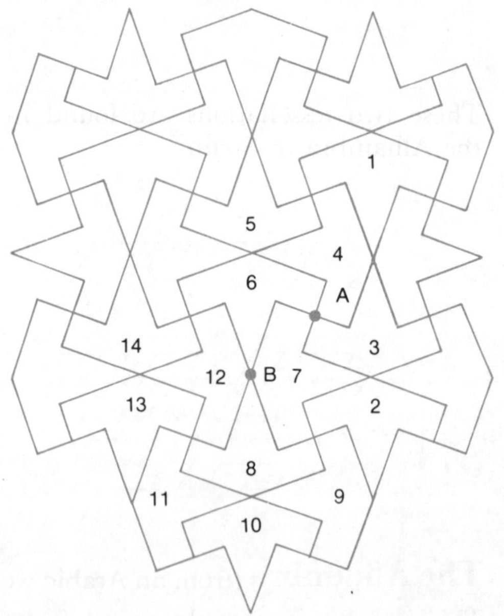

- Rotate 12 through 180° in a clockwise direction, about B. What is the numbered shape?
- Make up some similar questions of your own.

Tessellating Polyominoes

Draw diagrams on grid paper to show that *all* polyominoes with one, two, three, four or five squares will tessellate. Clearly label each set with the headings monomino, domino, trominoes, tetrominoes and pentominoes.
The twelve hexominoes will also tessellate. Outline them and colour, so that a shape does not have an edge in common with a shape of the same colour.

- Which heptomino (7 squares) will *not* tessellate?
- Can you form tessellations from all the triamonds? tetriamonds? pentiamonds?
- Which polyhedra (solids) can be arranged to fill space, without any gaps?

Tessellations with curved lines:

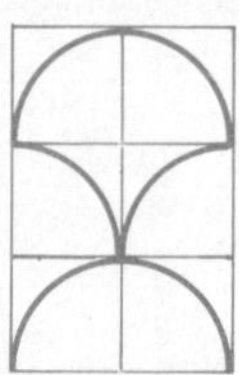

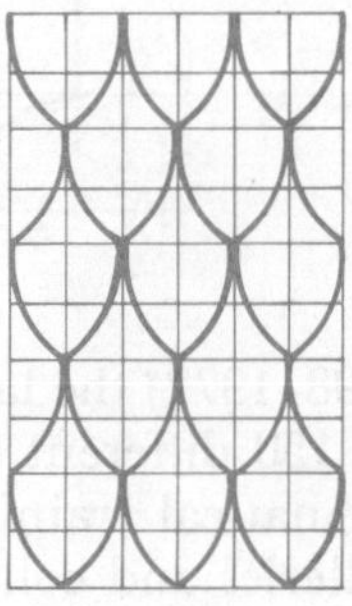

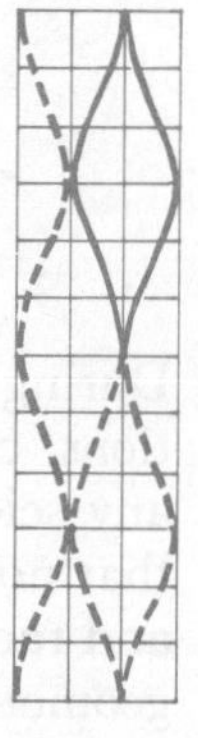

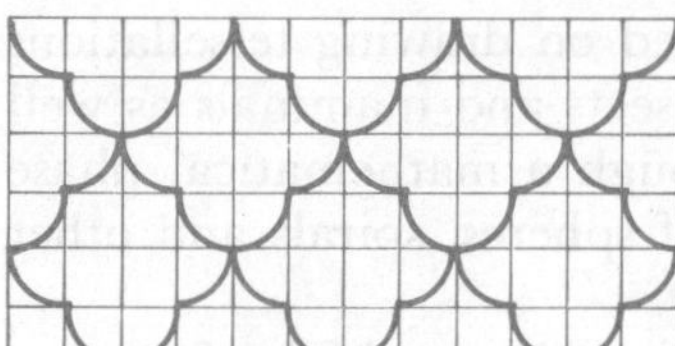

Note the similarity to the overlapping scales of a fish.

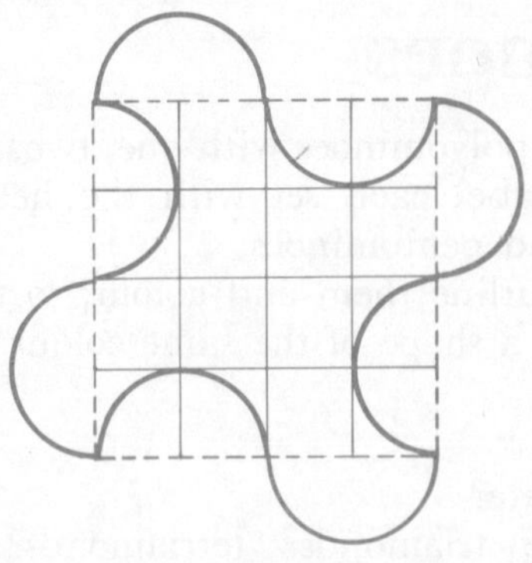

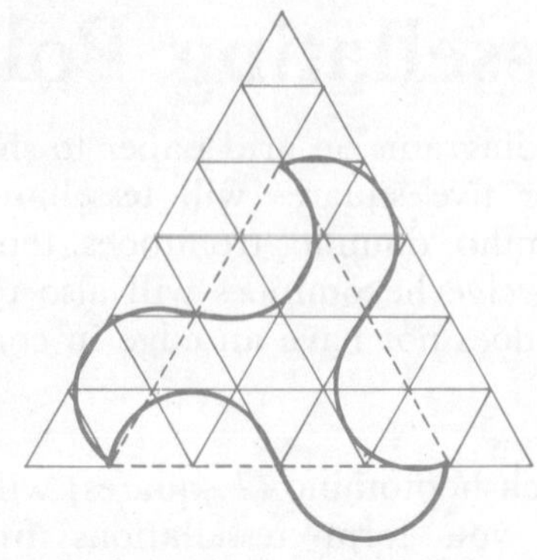

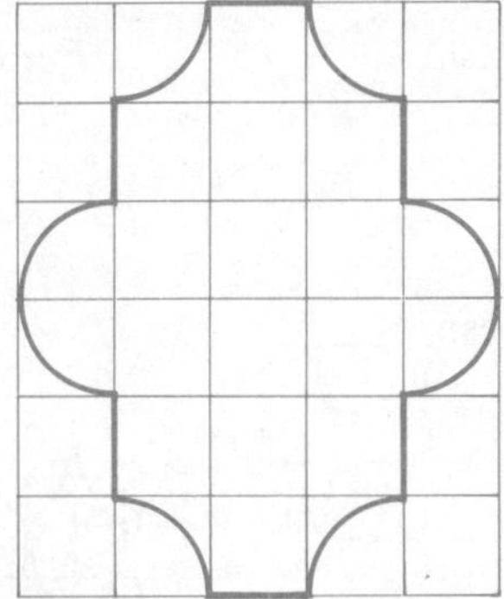

During his lifetime (1898–1972) the late M. C. Escher became a master of tessellations, creating at least 150 different ones in his early years without ever having any scientific or mathematical training! Altogether he created about 270 works that now appear in galleries and collections all over the world. He applied axial and radial symmetry, an unusual use of perspective, optical illusions, elementary geometrical principles, and his own fundamental, intuitive skills to create dynamic tessellations.

Discover some of his work in your library.

Unlike the Moors and Arabs, Escher concentrated on drawing tessellations of animate (living) creatures – birds, fish, reptiles, insects and mammals as well as scenes from Nature. Late in his life, he went through a mathematical phase when he used his talent for the unusual treatment of spheres, spirals and other shapes and solids.

Escher once said: "All my works are games. Serious games." The fact that his lithographs, woodcuts, wood engravings and mezzotints can be found hanging on the walls of both mathematicians' and scientists' homes proves that others also enjoy his "games" seriously.

Among crystallographers Escher is best known for his scores of ingenious mosaics on the plane, although strictly speaking they are not tessellations because the basic unit is not a polygon.

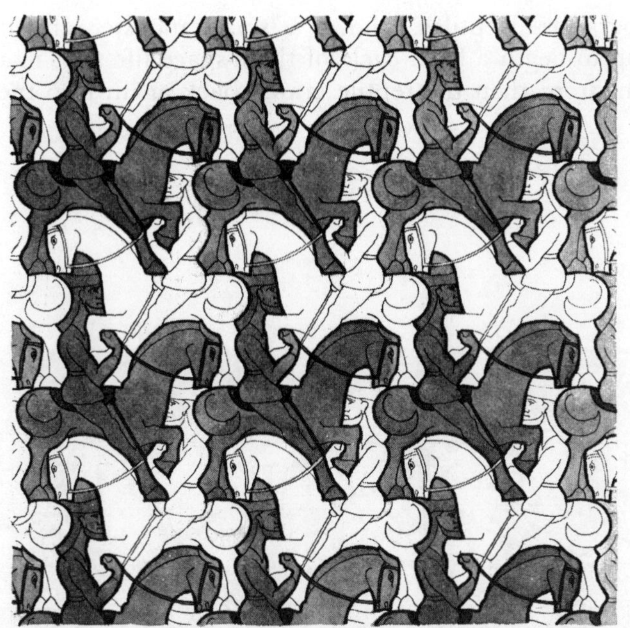

Symmetry Drawing, 1946, by M. C. Escher.

Sky and Water, 1938, by M. C. Escher.

This lithograph "Reptiles" shows a small monster crawling out of the hexagonal tiling to begin a brief cycle of three-space life that reaches its summit on the dodecahedron; then the reptile crawls back again into the lifeless plane.

Reptiles, 1943, by M. C. Escher.

Rep-Tiles

Only three regular polygons – the equilateral triangle, square and hexagon – can be used to tile a plane so that identical shapes are endlessly repeated. However, there is an infinite number of irregular polygons.
Consider this diagram:

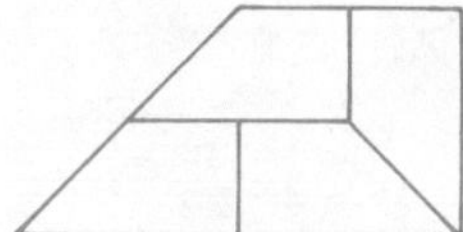

The trapezium has been divided into four smaller ones, each similar to the original, but congruent to each other.

Could it be divided into sixteen smaller ones?

Solomon W. Golomb invented these "replicating figures" or "rep-tiles". What geometrical shapes can you draw that could be called rep-tiles?

Are these shapes rep-tiles?
If so, draw diagrams in your book and dot in the boundaries.

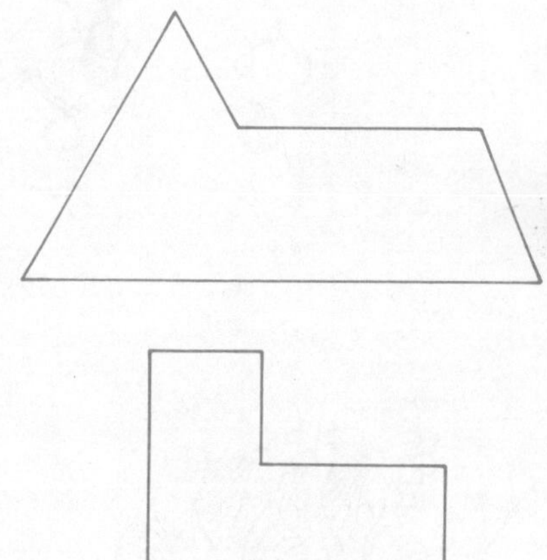

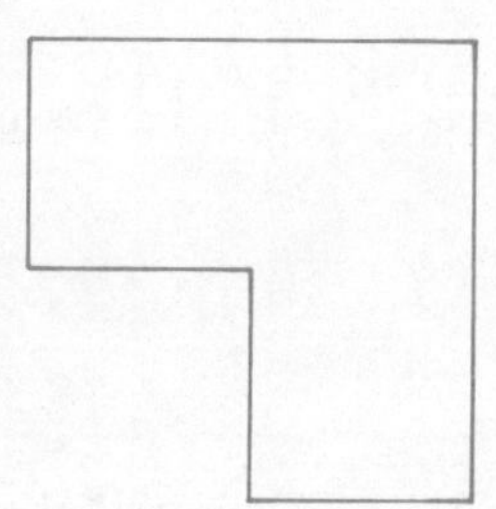

A Class Project

To make an attractive wall chart, the class can design or collect colourful tessellations like these.

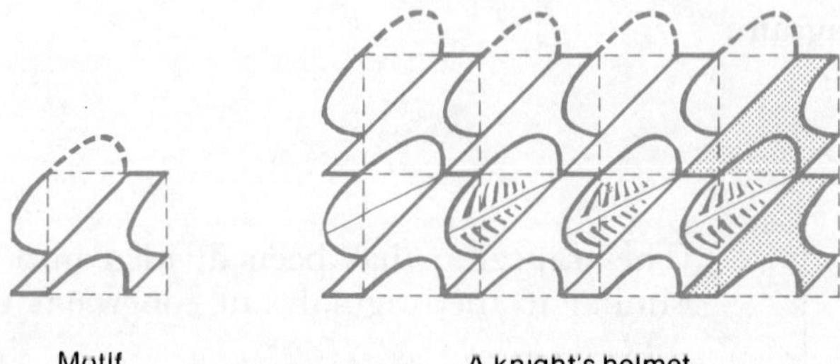

Motif A knight's helmet

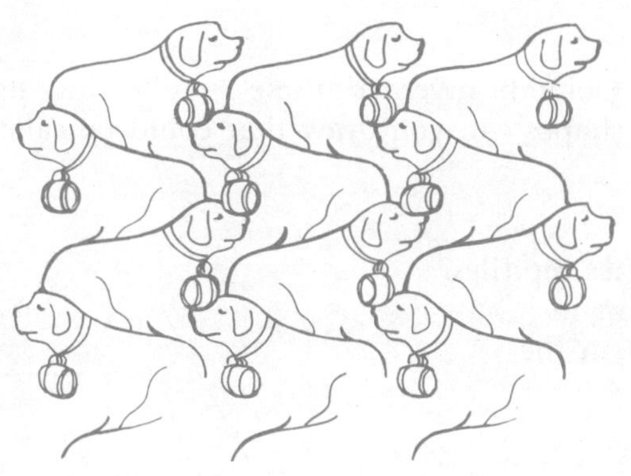

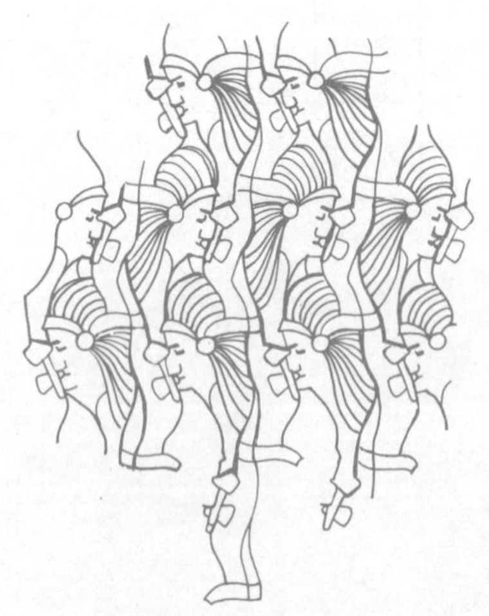

A sign of summer

A Challenge Problem

Cut out of cardboard at least eight copies of the first polygon and four copies of each of the other six. (Each differs from a square by four notches or "prongs".)

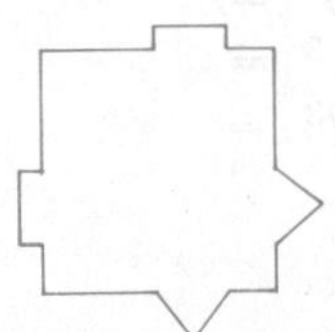

- Fit four together in the form of a 2×2 square. (You can turn them over or rotate them.)
- Fit six together to form a 2×3 rectangle.
- Fit nine together to make a 3×3 square.
- Fit sixteen together to form a 4×4 square.

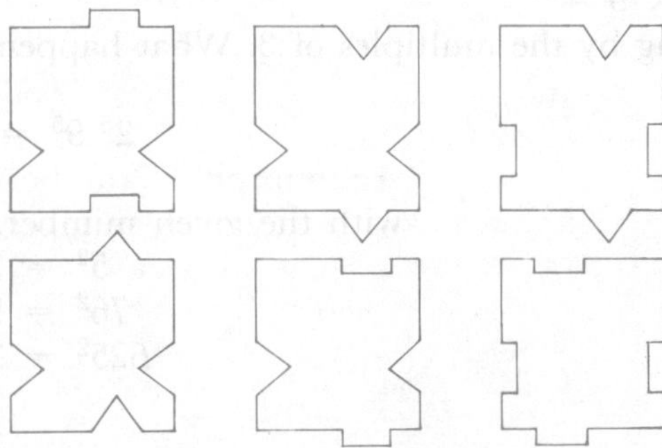

Unit 8 Number Relationships

Number Patterns

1. Did you know that:

$$1 + 2 = 3$$
$$4 + 5 + 6 = 7 + 8$$
$$9 + 10 + 11 + 12 = 13 + 14 + 15$$
$$16 + 17 + 18 + 19 + 20 = 21 + 22 + 23 + 24 \text{ ?}$$

2. $13^2 = 169$
   $31^2 =$
3. $\dfrac{666}{\cdot 666} =$ $\quad\dfrac{777}{\cdot 7} - \dfrac{77}{\cdot 7} =$ $\quad 999 + \dfrac{999}{999} =$
4. Is eleven thousand eleven hundred and eleven divisible by three?
5. The square root of any number n is always smaller than n. True or false?
6. Write the digits 9 to 1, in order, backwards.
7. Continue the next two lines in these patterns:

$9 \times 1 + 1 =$	$9 \times 0 + 1 =$	$9 \times 0 + 1 =$
$9 \times 2 + 2 =$	$9 \times 1 + 2 =$	$9 \times 1 + 2 =$
$9 \times 3 + 3 =$	$9 \times 2 + 3 =$	$9 \times 12 + 3 =$
		$9 \times 123 + 4 =$

$8 \times 1 + 1 =$
$8 \times 12 + 2 =$
$8 \times 123 + 3 =$
$8 \times 1234 + 4 =$

Narcissistic Numbers

$1^3 + 5^3 + 3^3 =$
$3^3 + 7^3 + 0^3 =$
$3^3 + 7^3 + 1^3 =$
$4^3 + 0^3 + 7^3 =$

$12\,345\,679 \times 9 =$
$12\,345\,679 \times 18 =$

$037\,037\,037\,037 \times 3 =$
$037\,037\,037\,037 \times 6 =$
$037\,037\,037\,037 \times 9 =$

Continue multiplying by the multiples of 3. What happens then?

$9\,109 \times 1 =$
$\times 2 =$
$\times 3 =$
$\times 4 =$
$\times 5 =$
$\times 6 =$
$\times 7 =$
$\times 9 =$

$2^5\, 9^5 = 2\,595$

Automorphic Numbers when squared, end with the given number.

$5^2 = 25$
$76^2 = 5\,776$
$625^2 = 390\,625$

$$1 \times 1 = 1$$
$$11 \times 11 = 1\,2\,1$$
$$111 \times 111 = 1\,2\,3\,2\,1$$
$$1\,111 \times 1\,111 = 1\,2\,3\,4\,3\,2\,1$$
$$11\,111 \times 11\,111 = 1\,2\,3\,4\,5\,4\,3\,2\,1$$

Continue the pattern until you multiply 111 111 111 by 111 111 111.

$$1\,0\,8\,9 \times 9 = 9\,8\,0\,1$$
$$1\,0\,9\,8\,9 \times 9 = 9\,8\,9\,0\,1$$
$$1\,0\,9\,9\,8\,9 \times 9 = 9\,8\,9\,9\,0\,1$$
$$1\,0\,9\,9\,9\,8\,9 \times 9 = 9\,8\,9\,9\,9\,0\,1$$

Add two more lines to this pattern.

Odd Numbers

```
1
3   5
7   9   11
13  15  17  19
21  23  25  27  29
31  33  35  37  39  41  . . .
```

Add the lines horizontally.

```
1
8
27
. . .
. . .
```

What type of numbers do you get in your totals?
Can you predict the sum of the digits in the twelfth row?

Amicable Pairs

$3\,869 = 62^2 + 05^2$ and $6\,205 = 38^2 + 69^2$
$5\,965 = 77^2 + 06^2$ and $7\,706 = 59^2 + 65^2$

Cyclic Numbers

$\frac{1}{7}$ as a decimal is ?
Now multiply 14 2857 $\times 1 =$
$\times 2 =$
$\times 3 =$
$\times 4 =$
$\times 5 =$
$\times 6 =$
$\times 7 =$

- What is the sum of 142 and 857?
- Add the first and fourth answers.
- Add the second and fifth.
- Add the third and sixth.
- Calculate the sum of the first six answers.

0 7 6 9 2 3	$\times$ 1	=	0 7 6 9 2 3
	$\times$ 10	=	7 6 9 2 3 0
	$\times$ 9	=	6 9 2 3 0 7
	$\times$ 12	=	9 2 3 0 7 6
	$\times$ 3	=	2 3 0 7 6 9
	$\times$ 4	=	3 0 7 6 9 2

Note that the first column of numbers after the equal signs reads 076923 vertically and all the other columns have the same sequence of digits as do the rows.

If this magic number is now multiplied by 2, 7, 5, 11, 6 and 8 respectively, we get a new number, 153846 – that is twice 076923 and has the same properties.

Note that the sum of the digits in each row and column in all these numbers adds up to 27.

7 6 9 2 3	$\times$ 2	=	1 5 3 8 4 6
	$\times$ 7	=	5 3 8 4 6 1
	$\times$ 5	=	3 8 4 6 1 5
	$\times$ 11	=	8 4 6 1 5 3
	$\times$ 6	=	4 6 1 5 3 8
	$\times$ 8	=	6 1 5 3 8 4

Interesting Patterns

999 999 × 2 =
× 3 =
× 4 =
× 5 =
× 6 =
× 7 =
× 8 =
× 9 =

Work out the remainder if:

2 519 ÷ 10 =
÷ 9 =
÷ 8 =
÷ 7 =
÷ 6 =
÷ 5 =
÷ 4 =
÷ 3 =
÷ 2 =
÷ 1 =

Complete 99 × 11 = 1089
99 × 22 = 2__8
99 × 33 = _26_
99 × 44 = 4_5_
99 × 55 = _4_5
99 × 66 = __3_
99 × 77 = ___3
99 × 88 = _7__
99 × 99 = ____ (Compare this with your first answer.)

Pairs of two digit numbers can have the same product when both numbers are reversed. For example:

12 × 42 = 21 × 24
12 × 84 = 21 × 48
13 × 62 = 31 × 26
23 × 96 = 32 × 69
24 × 63 = 42 × 36

Check to see that these sentences are true.

An Oddity

A constellation of six digits 2, 3, 7, 1, 5 and 6 has the interesting property that:

$$2 + 3 + 7 = 1 + 5 + 6$$

and

$$2^2 + 3^2 + 7^2 = 1^2 + 5^2 + 6^2. \quad \text{Correct?}$$

More than 200 years ago, Goldbach and Euler developed many equations that generate such constellations.

Calculating the Square Root – Chinese Method

This depends on the fact that the sum of successive odd numbers always produces a square number.

$$1 + 3 = 4$$
$$1 + 3 + 5 = 9$$
$$1 + 3 + 5 + 7 = 16\ldots$$

Follow this:

$\sqrt{16} =$	$16 -$	Step 1
	1	
	$15 -$	Step 2
	3	
	$12 -$	Step 3
	5	
	$7 -$	Step 4
	7	
	0	The answer is the number of steps involved after subtracting the odd number sequence.
$\therefore \sqrt{16} = 4.$		

Try $\sqrt{36}$ yourself.

What are triangular numbers? Write down the first five. Note that no triangular number ends in 2, 4, 7 or 9.

The ancient Greek mathematician Diophantus, who lived in the 3rd Century B.C. found a simple connection between triangular numbers (T) and square numbers (K):

$$8T + 1 = K$$

Test this formula with a selection of examples.

Decimal Patterns

$1/9 = 0\cdot$	$1/11 =$	$3/7 =$	$1/13 =$
$2/9 = 0\cdot$	$2/11 =$	$4/7 =$	$2/13 =$
$3/9 = 0\cdot$	$3/11 =$	$5/7 =$	$3/13 =$
		$6/7 =$	

The third group is said to be cyclic. Can you work out why?

SEQUENCES

A \ D	1	2	3	4	5
1	1	2	6	24	·
2	5	9	11	·	60
3	10	28	·	8	20
4	16	·	27	3	5
5	·	126	38	0	1

Fill in the blank squares. 1D, 2D, 3D, 4D, 5D and 1A are regular sequences of numbers whose methods of formation you have to discover. Rows 2A, 3A, 4A and 5A are obtained from the following formulae:

2A: n, $2n - 1$, $2n + 1$, $3n$, $12n$.
3A: ab, $4b + c$, $2b + c$, $4a$, $\frac{1}{2}bc$.
4A: a^2, $b^2 + 1$, $a + 3b - 1$, $\frac{1}{4}(a + b)$, $a + 1$.
5A: $x + y$, xy, $x + y + z$, $y - z + 1$, $z - y$.

What are the numerical values of the symbols used?

MIND READING EXERCISE REVEALED BY ALGEBRA

	Let x be the number.
Instructions: – Think of any number.	x
Multiply by 5.	$5x$
Add 6.	$5x + 6$
Multiply by 4.	$20x + 24$
Add 9.	$20x + 33$
Multiply by 5.	$100x + 165$
Subract 165.	$100x$

The answer is 100 times larger than the original number.

Limited Numbers

Study these examples:

$$1 = \frac{3 - 1}{2}$$

$$2 = \frac{3 + 1}{2}$$

$$3 = \frac{3! \times 1}{2}$$ (factorial 3 . . . 3 × 2 × 1)

$$4 = 3 + 2 - 1$$

$$5 = \frac{3 + 2}{1}$$

Now see if you can work out all the numbers 1 to 12, using only 4 ones. You can use all the operations illustrated in the example.

Multiplication Pattern

On grid paper, number the horizontal axis 0 to 60, the vertical axis 0 to 40.

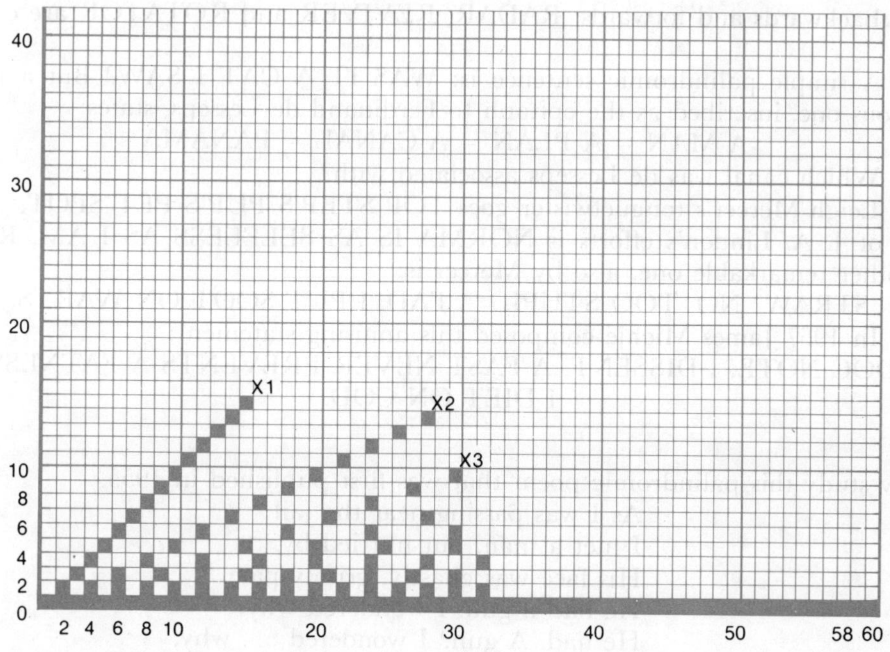

Instructions:

1. Continue colouring every second square on the horizontal row 2.
2. Continue to colour every third square in row 3.
3. Continue to colour every fourth square in row 4 and so on, until you fill the grid.

When complete, your graph will have many patterns. Can you see the patterns of the multiplication tables?

Use the graph to find:

- The product of 13 and 3.
- The factors of 48.
- Why 5 × 7 = 7 × 5 (What is this law?)
- Explain why the L.C.M. of 4 and 5 is 20.

Palindromes

A palindrome is a word, sentence or set of sentences that can be spelt the same way backwards and forwards. RADAR, REVIVER and ROTATOR are examples.

A simple palindromic sentence is: WAS IT A CAT I SAW? But a more famous one, inscribed as the epitaph to Ferdinand de Lesseps, states:

A MAN – A PLAN – A CANAL – PANAMA.

Which canal was de Lesseps associated with?

Leigh Mercer's tonguetwister goes TOP STEP'S PUP'S PET SPOT, while one of J. A. Lindon's efforts is NORMA IS AS SELFLESS AS I AM, RON. Another remarkable one, also by Mercer is:

STRAW? NO, TOO STUPID A FAD. I PUT SOOT ON WARTS.

In 1967 James Michie composed this amusing statement:

DOC NOTE, I DISSENT. A FAST NEVER PREVENTS A FATNESS. I DIET ON COD.

Now study this palindromic poem that was first published in 1955.

As I was passing near the jail
I met a man, but hurried by.
His face was ghastly, grimly pale.
He had a gun. I wondered why
He had. A gun? I wondered . . . why,
His face was ghastly! Grimly pale,
I met a man, but hurried by;
As I was passing near the jail.

J. A. Lindon.

Palindromic numbers, like words, are the same way backwards as forwards. In mathematics we also have dates such as 1991. Can you remember any palindromic numbers from Pascal's triangle? Here are some exercises that eventually give palindromic answers:

Add the reverse of any integer to the integer itself.

Example: 38 + 83 = 121 (a palindrome)

Many steps may be necessary:	168 +	
	861	**Reverse and add**
	1029 +	
	9201	**Reverse and add**
This has been done in three steps	10230 +	
	03201	**Reverse and add**
	13431	**Answer**

There are twenty-four steps in going from 89 to 8 813 200 023 188! There is only one number that has defied all attempts of both man and computers, after thousands of steps, to become palindromic. It is 196, the square of 14.

Now complete this table:

Number	Number of Steps	Palindrome
24	1	66
168	3	13,431
68		
192		
364		
553		
89		

- The palindromic number 111 is divisible by a small prime number. What is it? If 111 is cubed, the palindromic answer 1 367 631 results.
- Also, $(836)^2$ = 698 896 which is the smallest palindromic square with an even number of digits. Is is still palindromic if it is turned upside down?
- Suppose a car's meter showed 15 951 km (a palindrome). After travelling for two hours the driver noticed another palindromic number had come up. At what speed had the car been travelling?
- Palindromes have uses in other fields — melodies that are the same backwards, paintings and designs with reflective symmetry and so on.

Cryptics

The answers to the following clues are all mathematical words.
The number of letters in the answer are shown in parentheses after the clue.

1. Right leader is in the current of this curve. (3)
2. Ready to go for this group. (3)
3. A doctor of divinity can make additions. (3)
4. Initiate some useful methods to find the total. (3)
5. The course of the Nile is wrong. (4)
6. Automobile Association is about this region. (4)
7. The Italian is flattened by this weight. (4)
8. The lockup in the postscript to make the addition. (4)
9. There are no other factors to provide an untidy R.I.P. for me. (5)
10. The young journalist is in charge of the third degree. (5)
11. The answer is found in sum to be less. (5)
12. Enacts badly cutting the circle. (6)
13. The act is in for a part. (6)
14. This smaller group is sure to best us. (6)
15. In this area, initially, some quadrupeds usually avoided rowdy encounters. (6)
16. Hags can improve their lines with effective PR. (6)
17. 501, 6, and 500 to the east to factorise. (6)
18. Red cue is broken to make it smaller. (6)

The Golden Number

This is the number of any year in the Metonic lunar cycle (phases of the moon) of nineteen years, so named as important in fixing the date of Easter. A year's number in the cycle is called its golden number probably because of the colour with which it was marked in ancient calendars.

Golden numbers were introduced about the year 530 A.D. but were arranged as they would have been if they had been adopted at the time of the Council of Nicaea. The cycle was taken to begin in a year when the new moon fell on 25th January. This date for the new moon occurred in the year preceding the commencement of the Christian era and so, to find the golden number for any year the rule is:

Add one to the number of the year and divide the sum by 19. The quotient is the number of lunar cycles that have elapsed and the remainder is the golden number.

If the remainder is 0, the year is the nineteenth, or the last of the cycle.

The Golden Section

Geometry has two great treasures: one is the Theorem of Pythagoras; the other, the division of a line into extreme and mean ratio. The first we may compare to a measure of gold; the second we may name a precious jewel.

J. Kepler (1571–1630)

Fibonacci numbers appear to have had a strange influence on art and architecture. In describing the building of the Great Pyramid of Gizah about 4700 B.C, the papyrus of Ahmes reads: "The sacred quotient, seqt, was used in the proportions of our pyramids."

The Greeks called this sacred quotient the **golden section**. Legend has it that the mathematician Eudoxos (around 350 B.C.) was the first to try to find out why the golden section was so appealing. He worked out the mathematics of the golden proportion and found that it was a number that could be expressed as a formula! It was called **phi**, after Phidias, an artist who used the golden proportion in his sculptures.

Pythagoras suspected that the golden ratio was the basis for the proportions of the human figure. This he proved to be correct, showing that the human body is built with each part in a definite golden proportion to all the other parts. The height of a man to his navel height was calculated as approaching 1·618 . . .

Using the golden proportion frequently, the Greeks developed many excellent designs and patterns that were used extensively in pottery, ornaments, sculptures and paintings, as well as in architecture.

Known as the Law of Divine Proportion in medieval times, it was known to Dante, the Italian poet, who believed that "Nature is the art of God".

What is it?

The ratio between any two consecutive Fibonacci numbers (after 3) is approximately 1 : 1·6 (approaching 1 : 1·618). This is called the golden section or **Golden Ratio** and has intrigued people for centuries. It is the only number with a reciprocal exactly equal to one less than the number.

$$\therefore \quad 1/\Phi = \Phi - 1$$
$$\text{i.e.} \quad (\Phi)^2 - \Phi - 1 = 0$$
$$\therefore \quad \Phi = 1{\cdot}61803398\ldots$$

Although the ratio occurs on work in circles, pentagons and decagons, it is more notable in the **Golden Rectangle,** said to be one of the most visually satisfying of all geometric forms.

The golden proportion is a definite measurement. It is that division of any line in which the smaller interval is in the same proportion to the greater part as that greater part is to the whole.

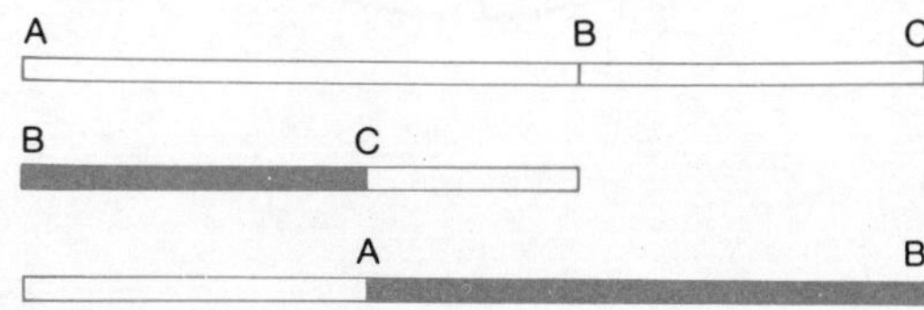

The golden section is at B when $\frac{BC}{AB} = \frac{AB}{AC}$

ACTIVITY

To find the golden section of a given line AB, draw a perpendicular at B and mark off along this perpendicular half the length of AB, at C.
Join AC, and with centre C cut AC at D so that CD also is half AB.

From A measure off AD, with compasses, then with centre A, radius AD, cut AB at E. AB is then divided at E such that:

$$\frac{EB}{AE} = \frac{AE}{AB} \doteqdot \frac{1}{1{\cdot}618}$$

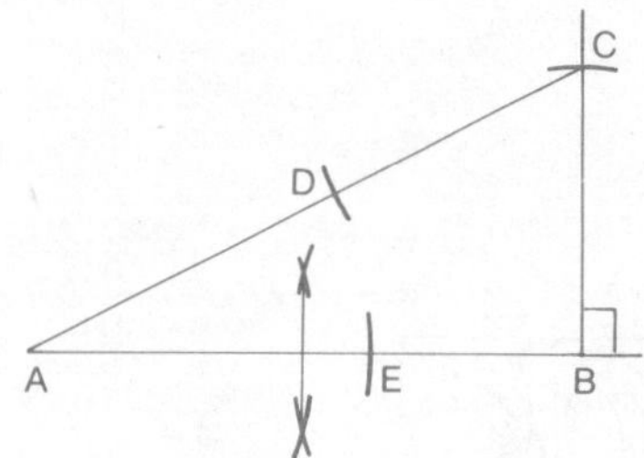

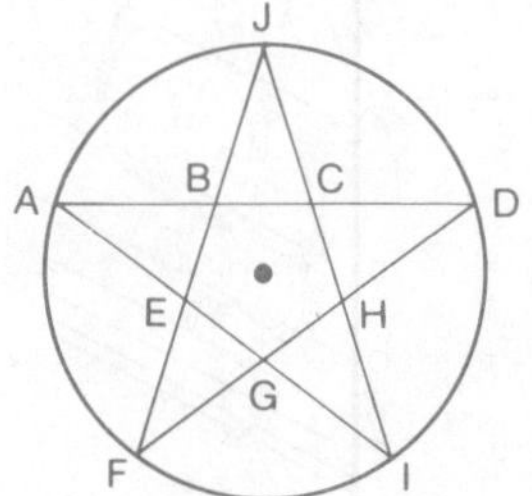

The various proportions of the lines on a pentagram inscribed within a circle are all exactly based on the golden ratio.

i.e. $\frac{BC}{AB} = \frac{AB}{AC} \quad \frac{CD}{AC} = \frac{AC}{AD}$ and so on.

Snowflake Analysis

$\frac{AB}{BC}$ = the golden proportion.

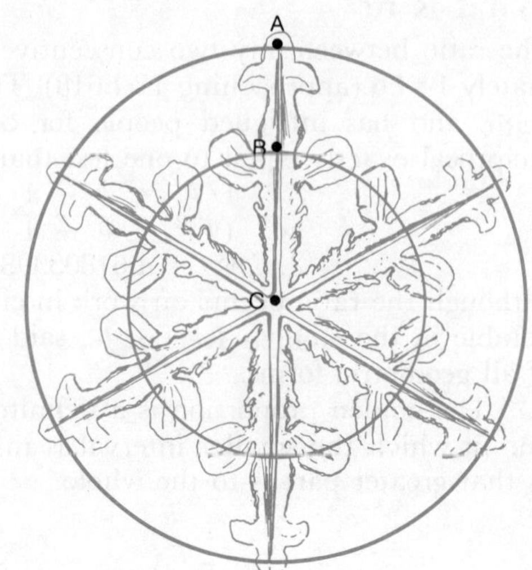

Butterfly Analysis

$\frac{AB}{BC}$ = the golden proportion.

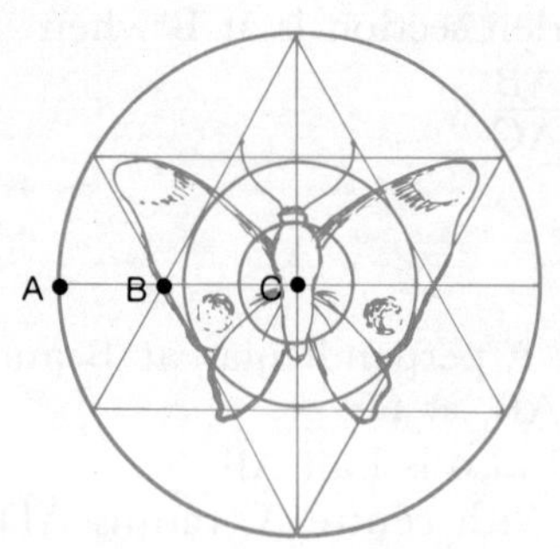

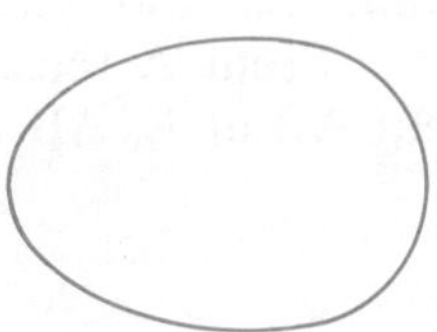

Egg Analysis

Dynamic symmetry because of the ratio.

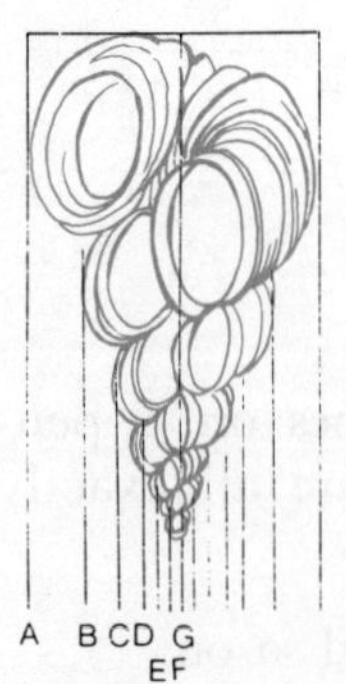

Shell Analyses

$\frac{BC}{AB} : \frac{AB}{AC} : \frac{CD}{BC} : \frac{BC}{BD}$ etc.

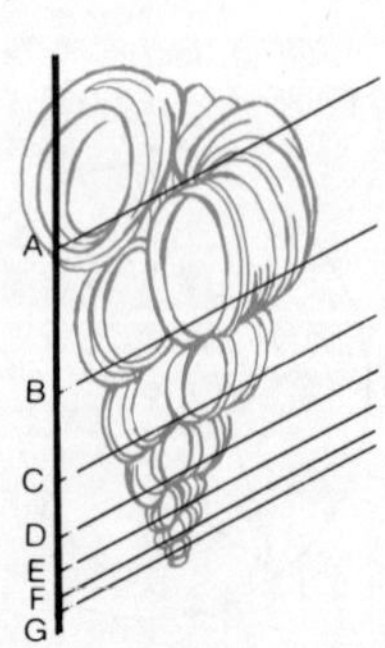

The Golden Rectangle

This is a rectangle that has the dimensions of length : breadth in the ratio of 1 : 1·62.

Look at these four rectangles and choose the one that is most appealing to you.

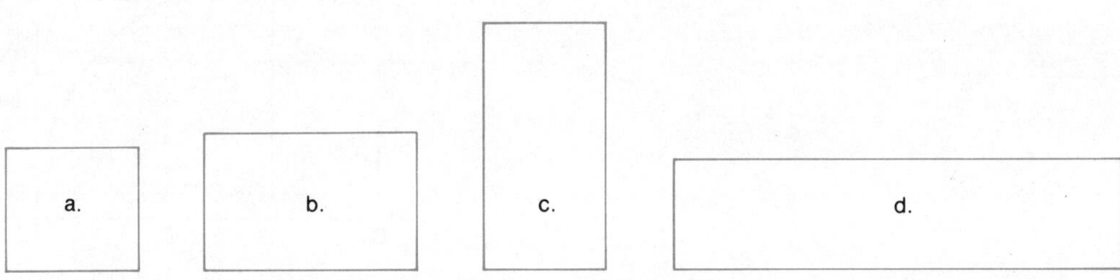

Make a survey in your class.

Measure the $\frac{\text{length}}{\text{breadth}}$ for each rectangle.

Which one is nearest to 1·62?
Which one is the **golden rectangle**?

Algebraically, $\frac{a}{b} = \frac{b}{a + b}$

Multiply both sides by $b(a + b)$

$$\therefore \quad a^2 + ab = b^2$$

or

$$b^2 - ab - a^2 = 0 \text{ (quadratic form)}$$

Divide both sides by a^2.

$$\therefore \quad \frac{b^2}{a^2} - \frac{b}{a} - 1 = 0$$

Substituting in the general quadratic formula

$$x. = \frac{-b \pm \sqrt{b^2 - 4ac}}{2a}$$

where $a = 1$

$b = -1$

$c = -1$

we get $\frac{b}{a} = \frac{1 + \sqrt{5}}{2}$ or $\frac{1 - \sqrt{5}}{2}$

i.e. $\frac{b}{a} = 1{\cdot}618034 \ldots$ or $-1{\cdot}618034 \ldots$

If you cut off a square from a golden rectangle the remaining piece is another golden rectangle. There is an area relationship involved between the two forms.

CONSTRUCTION OF THE GOLDEN RECTANGLE

1. Begin with any square ABCD.
2. Draw the vertical axis of symmetry, XY.
3. With centre Y, radius YB, draw an arc cutting DC, extended, at Z.
4. From Z draw a perpendicular ZE.
5. Extend AB to cut ZE at F.

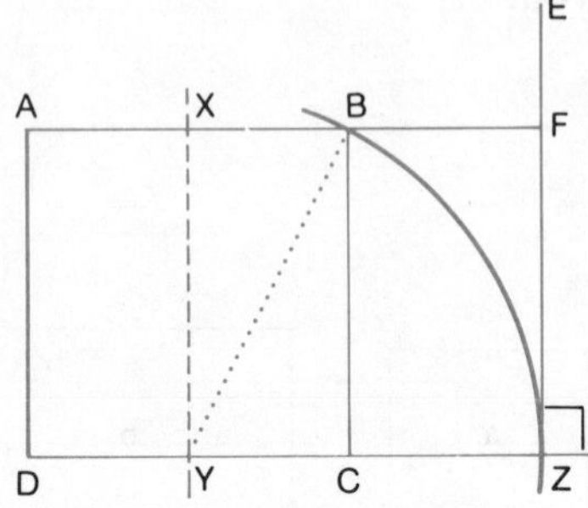

The Golden Rectangle is now AFZD. If the original square is taken away, the remainder will still be a golden rectangle!

In Leonardo da Vinci's drawing of an old man (probably himself) the artist overlaid the picture with a square, subdivided into rectangles, some of which approximate golden rectangles.

Leonardo also did many drawings of natural forms showing their mathematical proportions.

The unfinished "St Jerome" by da Vinci, painted about 1483, shows the great scholar with a lion lying at his feet. A golden rectangle fits so neatly around the figure that many experts believe da Vinci painted St Jerome to conform to these proportions.

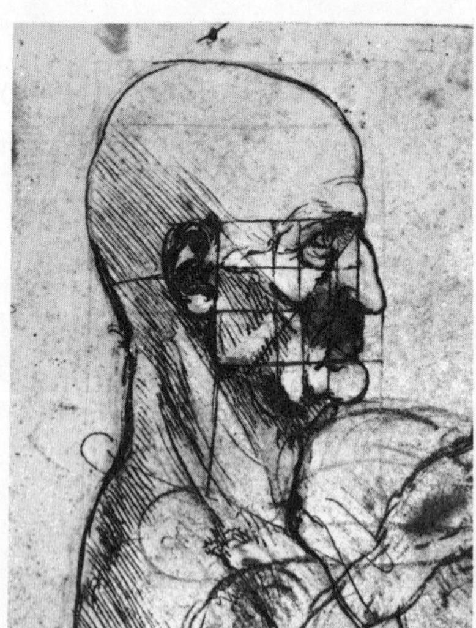

Study of Facial Proportions by Leonardo da Vinci.

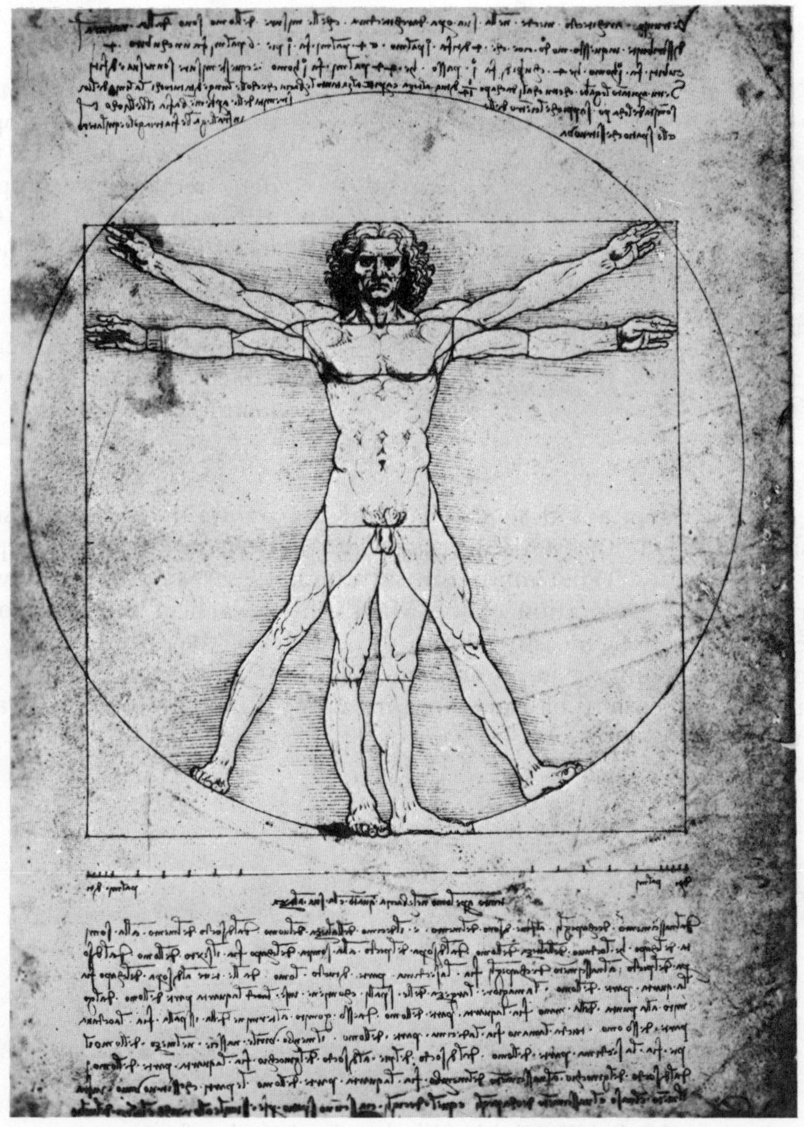

Study of Human Proportions is credited to Leonardo da Vinci but was probably drawn by one of his students named Vitruvius. It shows the analysis of the human figure and is drawn in accordance with mathematical laws.

Much later Dürer regarded the question of the "proper proportions" of the human figure as needing experimentation. He collected statistics on the human form then varied them according to a system of proportional increase and decrease until he produced figures too thin and too fat!

Architecture

The ancient temple, the Parthenon, on the Acropolis in Athens, was built in the golden rectangle's proportions, although in the 5th Century B.C. the builders probably did not know about it.

The temple was built to honour Pallas Athene, the fabled beauty, who was the Goddess of Wisdom.

ASSIGNMENTS

- How did the Egyptians design and build the pyramids with such skill and precision? Find out the statistics of the different pyramids and compare the base : height ratios. What mysticism surrounds these wonders of the world?
- Dürer's painting "Adoration of the Magi" shows each of the gifts positioned in the golden section, measured from the corners of the painting.
  Look up this painting in an art book.
- Measure the dimensions of a playing card to see if it is approximately a golden rectangle.

The Golden Triangle

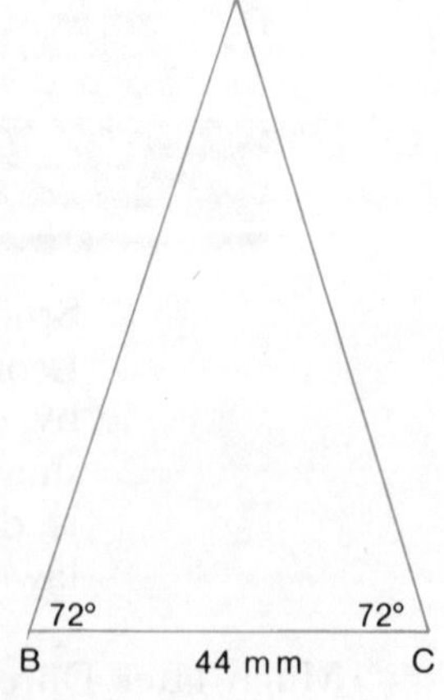

Measure the length of AB. Divide this by the length of the shortest side.
What do you get?

Make an exact copy of this in your exercise book.

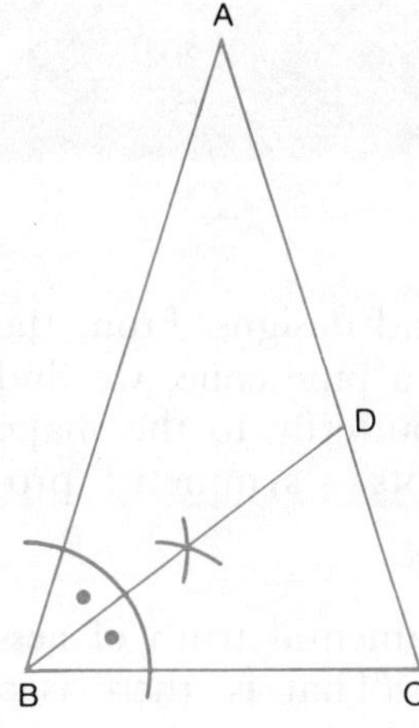

Bisect ∠ABC, using a protractor or compasses, so that the bisector cuts AC at D.

Work out the ratio $\frac{BC}{CD}$.

ABC is a golden triangle, and so is BCD. What do you think would happen if you bisected ∠BCD?

Continue this process of bisecting an angle in each of the new golden triangles formed, making sure you pick the angle in a corresponding position each time.

You will produce a set of "whirling" golden triangles, becoming so small you will not be able to draw them.

A FURTHER ACTIVITY

Make a similar triangle twice as large (enlargement factor of 2) on thick paper. Cut it out.

Starting from D, cut along the thick lines, bending each dotted line, in order, so that on the second fold D touches E (side AB). Continue in this way until the centre is reached.

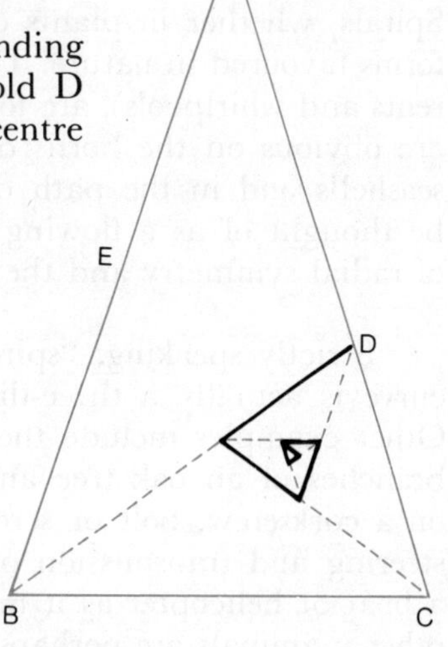

Now paste this folded triangle together to make a spiral model like a triangular snail shell. The golden ratio is actually found in real snail shells!

Unit 9 Mathematics in Nature

The world is one of geometrical curves, surfaces, shapes and designs. From the waves of the ocean, the shape of a galaxy, to the form of a pine cone, we find harmony of shape and design. From the symmetry of a butterfly to the shape of a snowflake, nature is a myriad of mathematical concepts – symmetry, proportion, perspective, ratios, patterns.

In the 13th Century, Thomas Aquinas stated a fundamental truth of aesthetics: "The senses delight in things duly proportional". (That is, there is a definite relationship between beauty and mathematics.)

These fundamental concepts are very important in the arts – music, painting, sculpture and architecture to name a few examples. Numerous shapes come to mind. Discuss some of them in class. For this section of the work, it would be advisable to keep a scrapbook, so that you can paste in (or draw) relevant pictures, patterns or illustrations.

Here are just a few examples for you to investigate:

Spirals

Spirals, whether in plants or animals, seem to be the most vital and dynamic forms favoured in nature. They swirl in the air (whirlwinds), roll in the sea (currents and whirlpools), are found in the home (water going down a plughole) and are obvious on the horns of sheep, on a snail shell (they indicate its age), on seashells and in the path of a gliding hawk. The basic spiral movement may be thought of as a flowing combination of two impulses: the circular impulse of radial symmetry and the flowing impulses of bilateral symmetry.

Strictly speaking, "spiral" refers to a plane (flat) curve. In most cases the curve is actually a three-dimensional curve called a **helix** (Latin for "snail"). Other examples include the growth of a stem tip upwards, the arrangement of branches of an oak tree and many man-made fabrications such as the thread on a corkscrew, bolt or screw, a carpenter's bit for a drill, helical gears in the steering and transmission of cars and the locus traced out by the propeller of a boat or helicopter as it moves along. The horns of rams, goats, antelopes and other mammals are perhaps more spectacular examples.

Would the locus of a merry-go-round operator, walking along the radius of the floor, at a constant speed, form a spiral or helix?

The Logarithmic Spiral

The logarithmic spiral was first recognised by Descartes, the man who invented co-ordinate geometry. This spiral intersects all the radii at a fixed angle and therefore is sometimes called an equiangular spiral. The distance of a logarithmic spiral from its pole (fixed point) increases in a geometrical sequence.

Logarithmic spirals occur often in nature, in the curve of an elephant's tusks and even on canaries' claws!

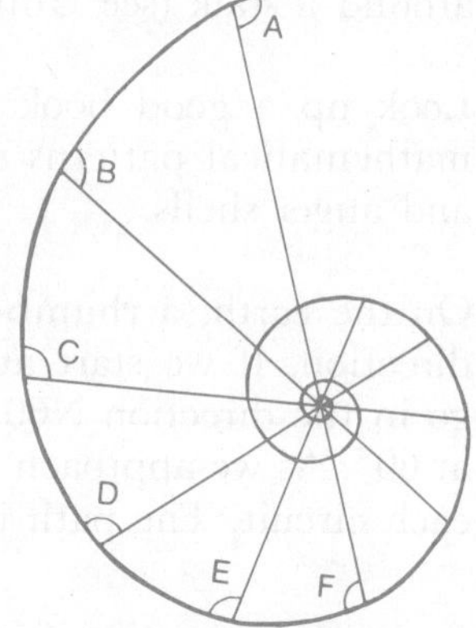

Jacob Bernoulli (1654–1705) found the logarithmic spiral so fascinating that he arranged to have it engraved on his tombstone with the Latin words *"Eadem mutata resurgo"* ("though changed I shall arise the same").

However, the Latin words were omitted and all the stonecutter could do was a crude spiral. It can be seen today at a place called Basel, in Switzerland.

Construction of a Spiral

The spiral of Archimedes is a geometrical interpretation of the square root of the numbers shown in the diagram with the distances of the spiral from its pole increasing in arithmetic sequence.

Using this as a guide, do the construction accurately in your book.

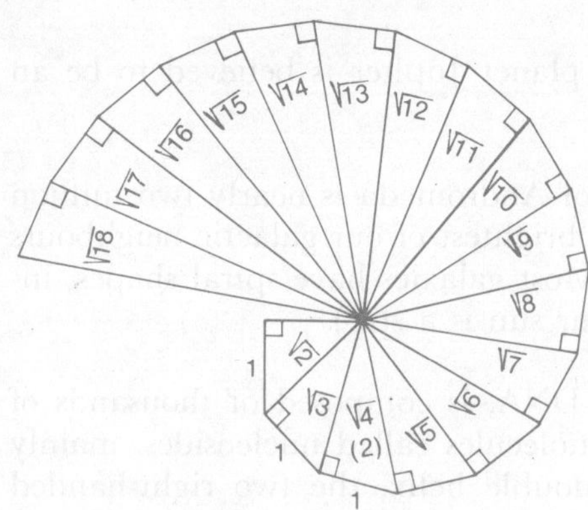

Check the length of $\sqrt{17}$ on the construction, then in your book of tables to see how accurate you are.

The human umbilical chord is a triple helix of one vein and two arteries that invariably coil to the left.

In Botany, helices are common in the structure of stalks, stems, tendrils, seeds, flowers, cones and leaves – even in the arrangement of leaves and branches around a stalk (see Unit 5 on Fibonacci numbers).

Look up a good book on seashells to find the infinitely delicate beauty and mathematical patterns of periwinkles, turban shells, abelones, tritons, top shells and auger shells.

On the earth, a rhumb-line is the path we travel if we keep going in the same direction. If we start at any place on the surface (except the North Pole) and go in the direction N60°E and continue in this direction, we cut each meridian at 60°. As we approach the North Pole, we spiral around it, getting a little closer each circuit. The path is an equiangular spiral.

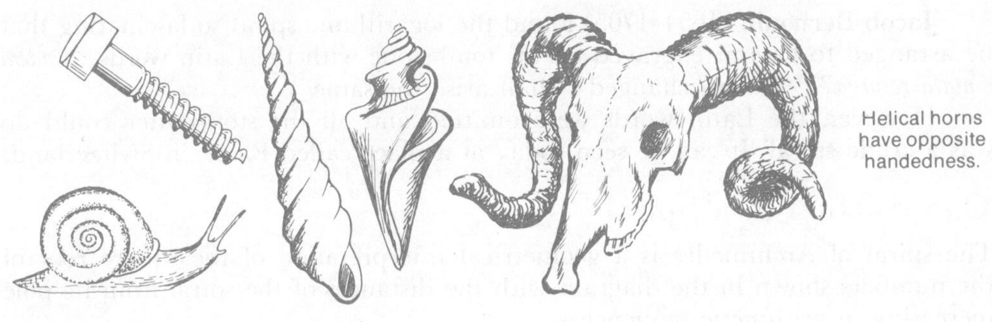

Did you know? The great Red spot on the planet Jupiter is believed to be an anticlockwise spiral cyclonic storm.

The great spiral galaxy in the constellation of Andromeda is nearly two million light years away. It is one of the nearest and brightest of our galactic neighbours and is easily seen by amateur astronomers. Most galaxies have spiral shapes, including the Milky Way (the one in which our sun is a star).

The substance that carries the code of life, DNA, is composed of thousands of units of only four varieties of quite simple molecules called nucleosides, mainly protein in nature, that are arranged in a double helix, the two right-handed helices twisting around each other.

M 101, Whirlpool Galaxy.

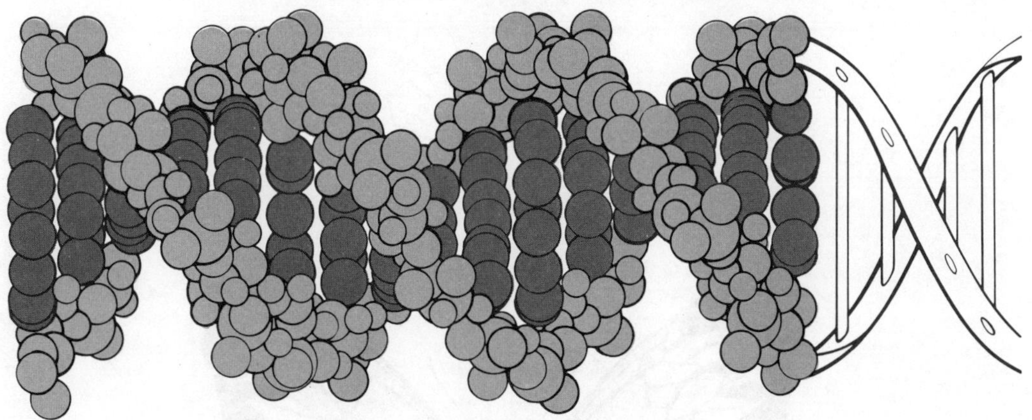

DNA arrangement.

The relationship between geometrical progressions and our ability to hear is very complex. In our ears there is an enormous number of tiny, sensitive receptors that pick up sounds and transmit nervous impulses to the brain, where they are interpreted as a particular sound. These receptors are hair-like forms graded in length to receive the whole range of resonant frequencies and they are arranged in a compact form in the **spiral cochlea** in our inner ears. The form of the geometrical number sequence is the link between conventional musical scales and the particular structure of the organ that receives these sounds.

- Have you ever connected the annual rings of tree trunks (that determine their ages) with concentric circles?
- What is the shape of the path traced out by planets around the sun? or comets? or satellites?
- Study a spider's web. What mathematical shape can you see? The transverse threads are parallel to one another and all are equally inclined to the main radial threads.
- What would be the path traced out by a fly walking out on the spoke of the wheel of a bike as it moves along at a constant speed?

Symmetry – The Mirror Image

Aristotle stated: "The chief forms of beauty are order and symmetry and precision which the mathematical sciences demonstrate in a special degree."

Radial symmetry (the basis being the circle, an ancient symbol of perfection) is found in many flowers and sea creatures, for example, in certain varieties of starfish.

Bilateral symmetry is found in bivalves, in many varieties of plants and animals (leaves, butterflies and so on) and many insects. You should be able to find plenty of illustrations of this.

Fruits also illustrate the regularity of natural shapes.

Cross section of an apple.

Cut section of an apple.

The apple is called a "false" fruit. Another is the strawberry. Try to find out the reason.

The word "composite" has application to fruit, too. The pineapple is an example because it is formed by the fusion of many flowers.

CONSTRUCTIONS

Draw a circle, radius 40 mm.
Divide the circumference into 24 equal parts.
Number the points.
Join 1 to 2, 2 to 4, 3 to 6, 4 to 8 and so on, doubling each number, until you come back to where you started.

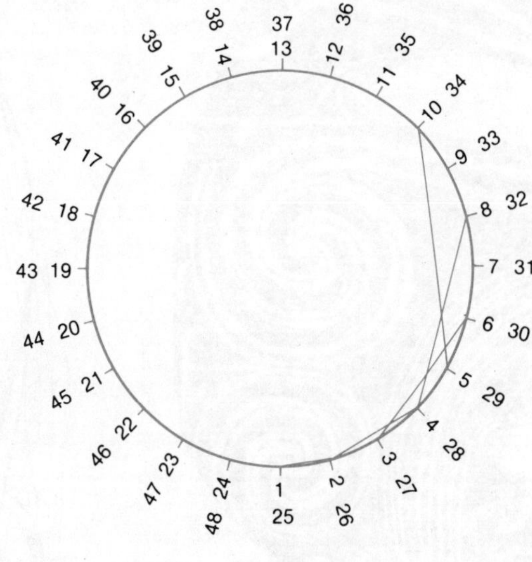

Note the similarity with the shape of the apple.

Describe another circle and mark it off in a similar way. This time, join points 1 to 3, 2 to 6, 3 to 9 and so on. What shape is formed?

Asymmetrical forms that appeal to us are said to have dynamic symmetry which, to an artist, is the most satisfying type.

Is the contour of a hen's egg more exciting than that of a marble? (The latter is called static symmetry.) Why?

ACTIVITIES

Organise a class competition to make a full-sized symmetrical mask. It could be made from cardboard, papier-mâché, or even clay. Refer to books on New Guinea or Africa.

Collect coloured pictures of asymmetrical objects and paste them on a wall chart.

Polygons

A polygon is a multi-sided figure. The sum of the sides is its perimeter. If the units forming the perimeter are all equal, the figure is said to be regular.

What is the smallest number of sides a polygon can have?

As the number of sides and angles increases, a transformation of the straight-sided figure takes place: it becomes more and more like a circle! Could a circle be called a regular polygon with an infinite number of sides?

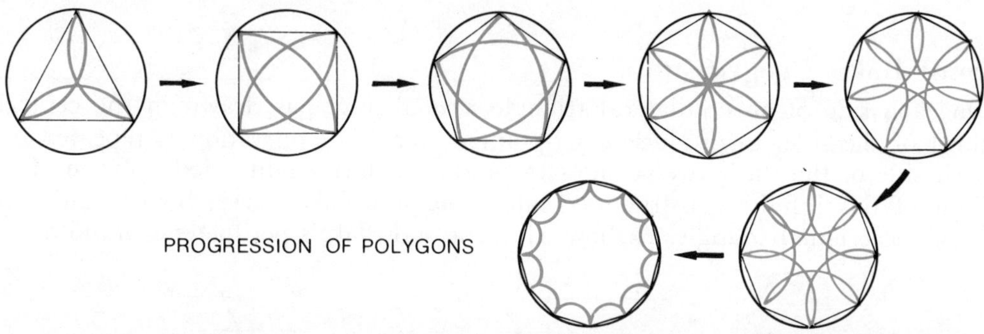

PROGRESSION OF POLYGONS

With your compasses, draw an *accurate* set of polygons, inscribed in circles, from a triangle to an octagon.
(*Hint:* work out the size of the angles at the centres of the circles.) Label each figure clearly.

Snow, like rain, is born from water – molecules of vapour flow freely through the air. When the temperature of air cools sufficiently, the movement of these vapour molecules is slowed down and together with the mutually attractive power all molecules possess, they unite and become visible as fog or clouds. If the temperature falls to freezing point they form crystals, the essence of symmetry and geometry.

The variety of snowflake shapes is inexhaustible, but most often follow the basic regular hexagon shape. Ice crystals are really hexagonal prisms constructed with geometrical precision.

An Englishman named Wilson Bentley photographed snow for over forty-six years and was unable to discover two identical shapes in over 5300 photographs!

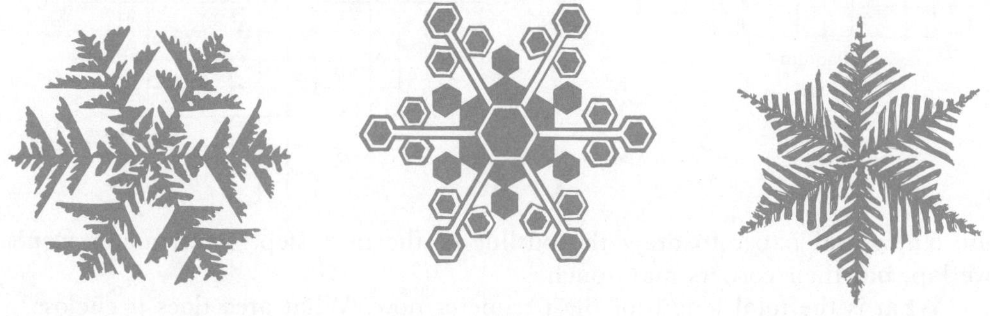

 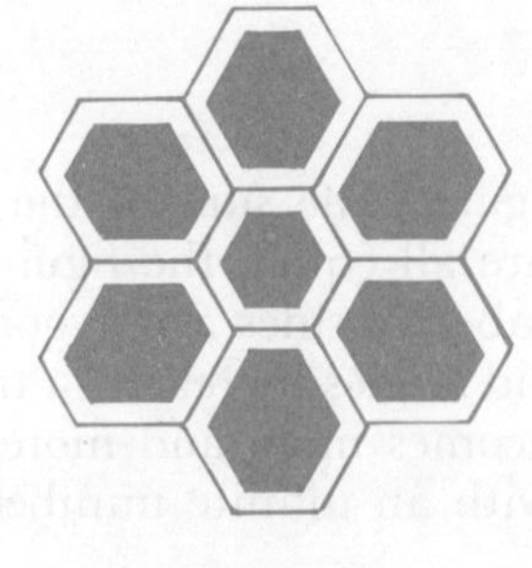 

Snowflake Progression

On each side of an equilateral triangle a smaller one is drawn in the central third of the side, to produce a six-pointed star. This operation is repeated on each side of the star's twelve sides to produce a forty-eight sided polygon. The limit of this infinite construction, called the snowflake curve, bounds an area $^8/_5$ of the original triangle. It shows that the snowflake's perimeter is infinite!

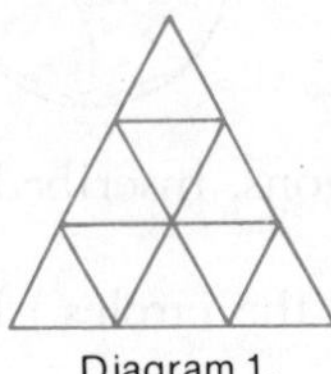

Diagram 1.

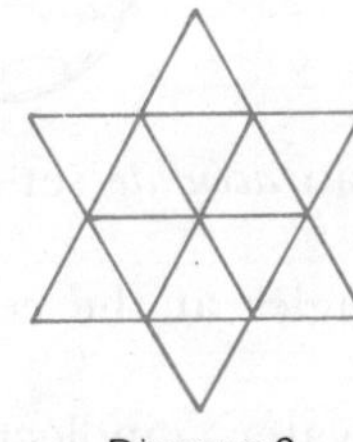

Diagram 2.

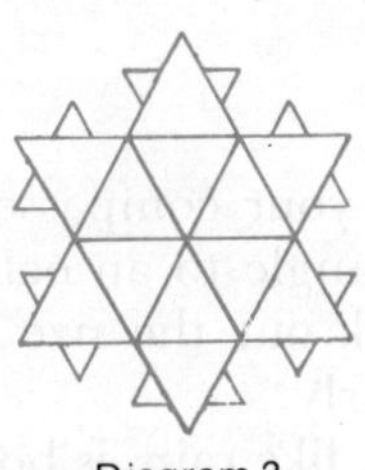

Diagram 3.

ACTIVITY

On the central third of each side of a unit square, draw four smaller squares (see Diagrams 4 and 5).

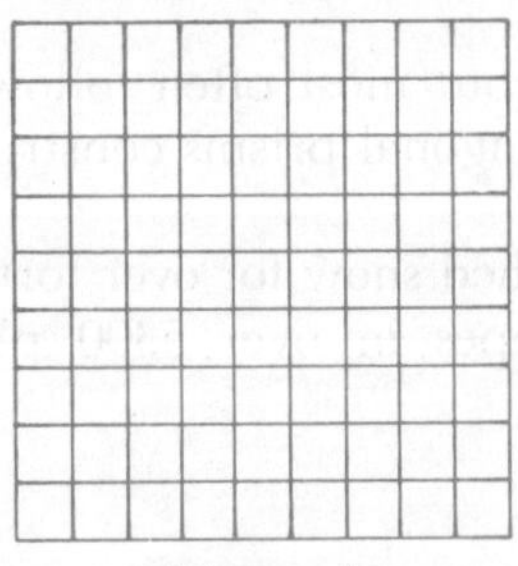

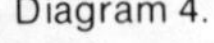

Diagram 4.

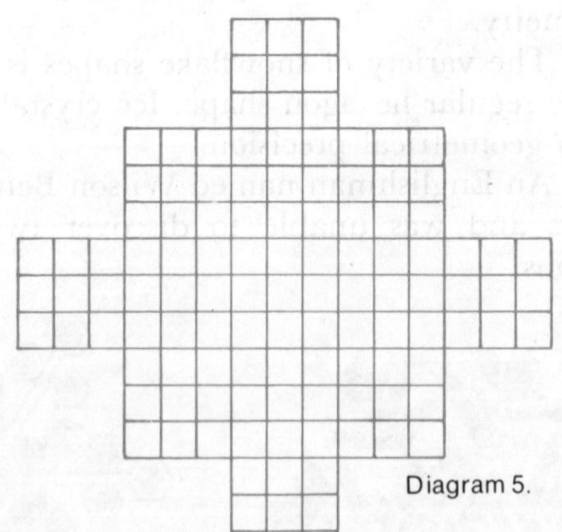

Diagram 5.

Use 5 mm grid paper to draw the outline of the next step. The squares won't overlap, but their corners may touch.

What is the total length of the perimeter now? What area does it enclose?

Polygonal forms occur often in nature, mostly in inanimate objects and in the lower orders of life – plants, insects and reptiles. By character the polygon is static. In animals it is found in many scale formations on skins which are shed, in the eyes of some insects (flies, bees, dragonflies in particular) where the many facets, varying from 10 to 30 000, are mainly hexagonal, forming a honeycomb-like mosaic. Each picture consists of a series of dots, like a printed newspaper picture.

Light visible to humans lies between red and violet. Though many insects are not so sensitive to the red end of the spectrum, they can see well into the ultra-violet, (which we cannot see). Honeybees are even able to see the plane of vibration of polarised light, a faculty they are able to use as a means of direction finding.

Many insects can distinguish colours, as is demonstrated by bees "going for" blue.

Did you know that a bee's wings make a figure-8 movement as they are raised and lowered?

A honeybee's dance also follows this shape.

Solids

Anyone who works with a microscope knows what a phenomenal world of shape, colour and variety there is to be seen through it.

The whole field of crystal structure (crystallography) mathematically developed in the 19th Century by such men as Bravais and Fedorov, was further explored by Von Laue and Braggs around 1912–1913, using diffraction of X-rays to reveal and measure the symmetries in the arrangement of planes of the atoms in crystals.

Most gemstones occur as well-formed crystals of varying shapes and sizes. A crystal is a solid body bounded by plane surfaces which are the external expression of a geometrical, internal arrangement of atoms. The angles made by corresponding faces of the crystals of any mineral are always the same, regardless of size. This regular arrangement gives six basic groups of crystal systems which differ from each other in their symmetry.

Isometric or Cubic System

This group has three crystallographic axes, all equal in length, at right angles to each other (called *a*-axes). *Examples:* halite, galena, fluorite, diamond, garnet, pyrite.

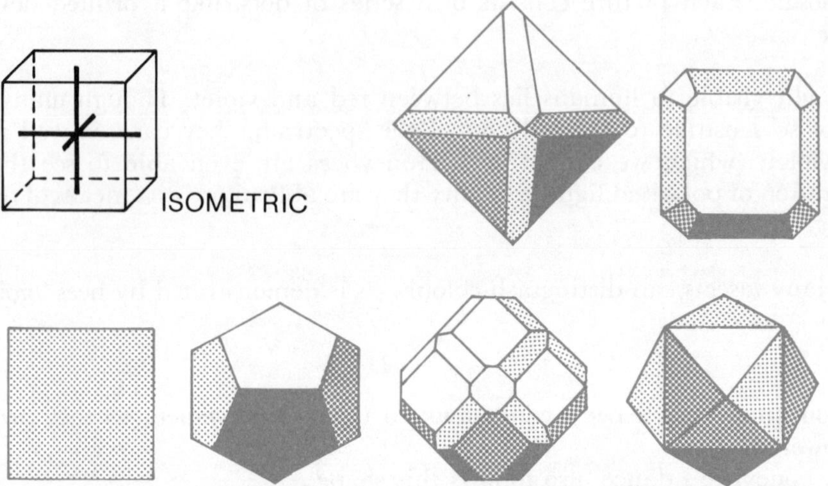

Tetragonal

In this group only the two horizontal axes are equal in length (*a*-axes), while the third one, the vertical axis, may be longer or shorter (*c*-axis). **Examples:** zircon, rutile, tetrahedrite (a copper mineral).

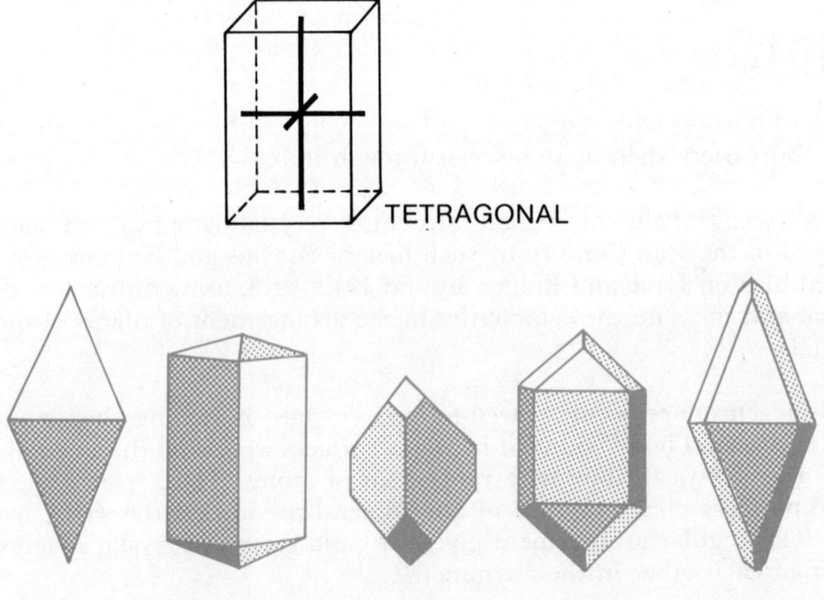

Orthorhombic

All three axes in these structures are unequal in length (*a* and *b*-axes for those in the horizontal plane and the *c*-axis for that in the vertical plane. **Example:** topaz.

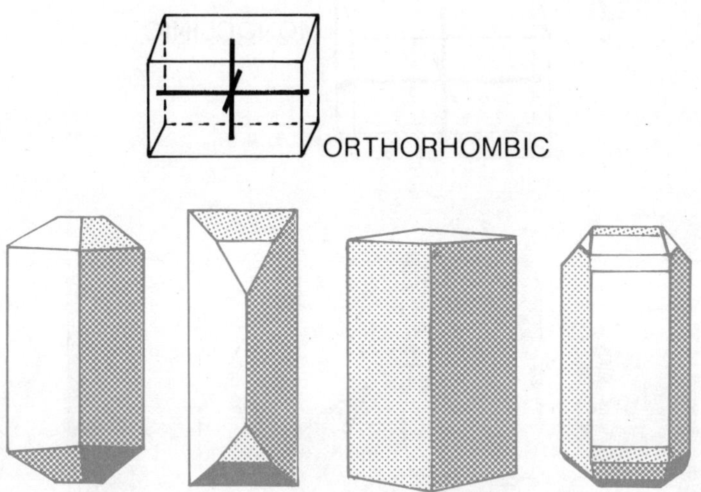

Hexagonal

This system has three horizontal axes of equal length (all *a*-axes) and a vertical *c*-axis. **Examples:** quartz, calcite, apatite, beryl.

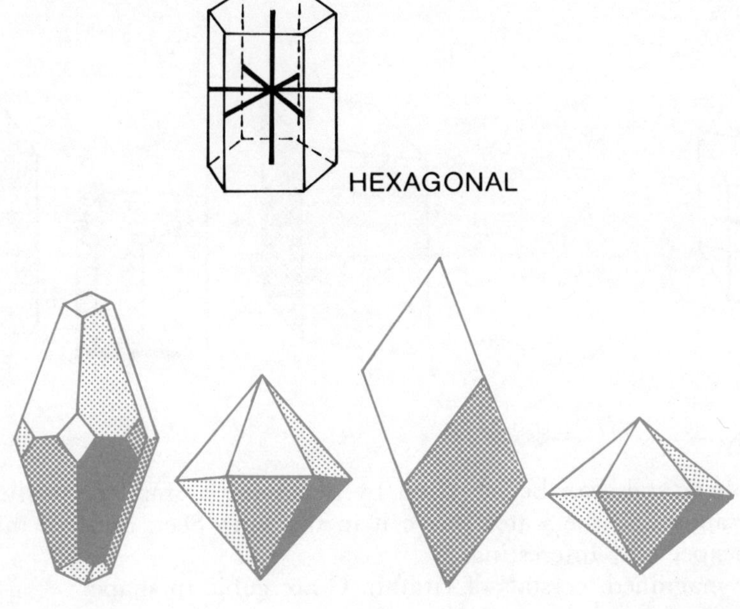

Monoclinic

All three axes are unequal with two intersecting at an oblique angle and the third being perpendicular to them.

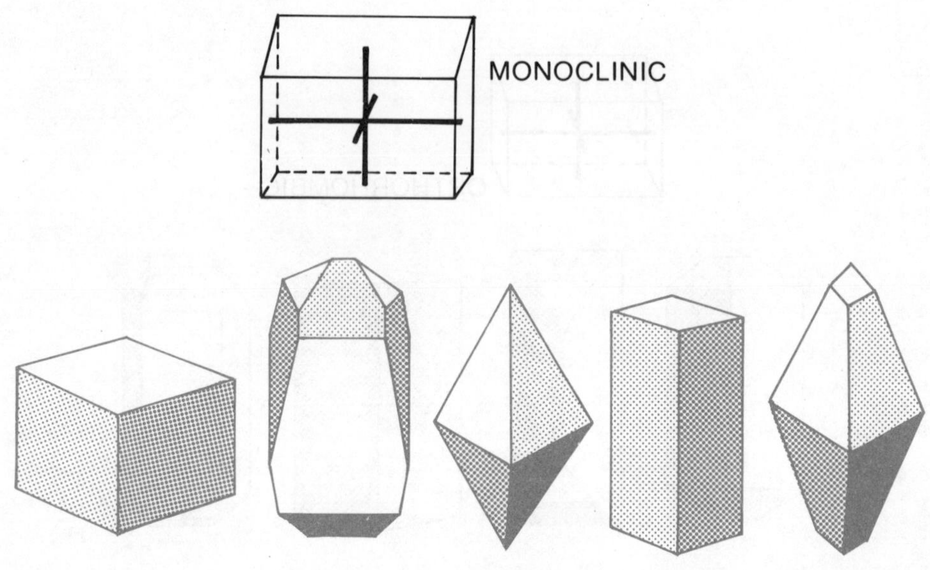

Triclinic

Three axes are unequal and all intersect at oblique angles. **Examples:** amazon stone, rhodonite.

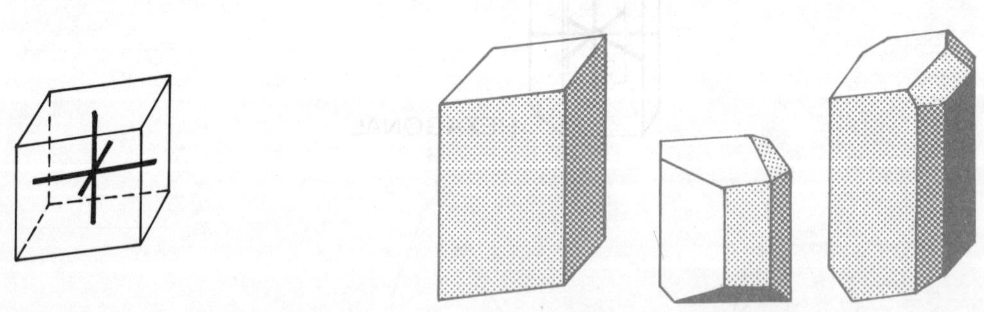

ACTIVITY

Sugar or salt crystals can be prepared by dissolving a small quantity in water and then evaporating the water (leave it in the sun). Seen under a microscope, they are unexpectedly interesting.

Highly magnified, crystals of vitamin C are cubic in shape!

Some of the most striking examples of regular polyhedra are found in minute sea creatures called radiolarians. Their skeletons are perfectly symmetrical polyhedra!

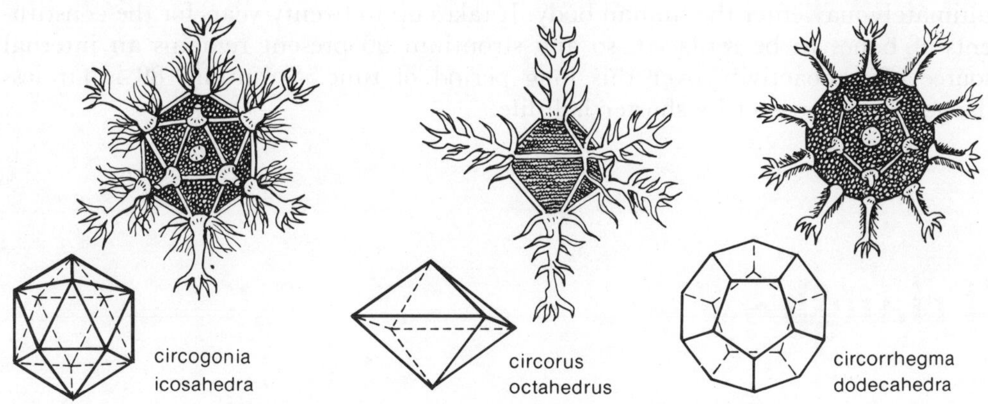

Sequences

In sequences such as $^1/_2$, $^1/_4$, $^1/_8$, $^1/_{16}$, $^1/_{32}$ . . . or 1, 0·9, 0·81, 0·729 . . . each term is obtained from the previous one by multiplying by the same factor, less than 1, and the terms of the sequences thus decrease in size very quickly.

Will the terms ever vanish if taken far enough?

Radioactive decay provides an example of a sequence like this. During the course of a year, a proportion of the atoms of a given piece of radioactive material disintegrate, emitting rays and leaving a disintegration product of less atomic weight. For any radioactive element, the proportion of atoms that disintegrate each year is **constant.** If, for example $^1/_{10}$ disintegrates each year, $^9/_{10}$ will remain unchanged. For such an element, the unchanged amount at the end of successive years would be:

Time (in years)	0	1	2	3	4	5	6	7	. . .
Mass unchanged	1	0·9	0·81	0·729	0·656	0·590	0·531	0·478	. . .

After six years a little more than half and after seven years, a little less than half the original material remains unchanged. Therefore half of it will have decayed in about six-and-a-half years. This period is called its **half-life.**

Would it be possible to calculate its whole life? Normally, only the half-life is calculated.

For radium $^1/_{2280}$ disintegrates each year, so the fraction $^{2279}/_{2280}$ has to be multiplied by itself until the answer is 0·5, in order to find its half-life, which is 1580 years! For uranium it is 4 600 000 000 years.

Strontium is an element that resembles calcium in some of its chemical properties and, like calcium, when some strontium is included in the diet, it goes into bone tissues. Several isotopes of strontium exist, two being radioactive –

strontium 89 with a half-life of 53 days and strontium 90 with a half-life of 19·9 years.

These are produced in thermo-nuclear explosions and are deposited in the fall-out on pasture land. This contaminated grass may be ingested by cows and ultimately may enter the human body. It takes up to twenty years for the constituents of bones to be replaced, so any strontium 90 present remains an internal source of radioactivity over this long period of time. Strontium 89 is far less dangerous because of its shorter half-life.

Triangles

Many of nature's most pleasing structures are enclosed within the sides of a triangle. The wings of insects, fins of fish and even the internal strutting in the wing bone of a vulture (as with most birds) shows the triangular arrangement, that is similar in construction to an aeroplane's wing.

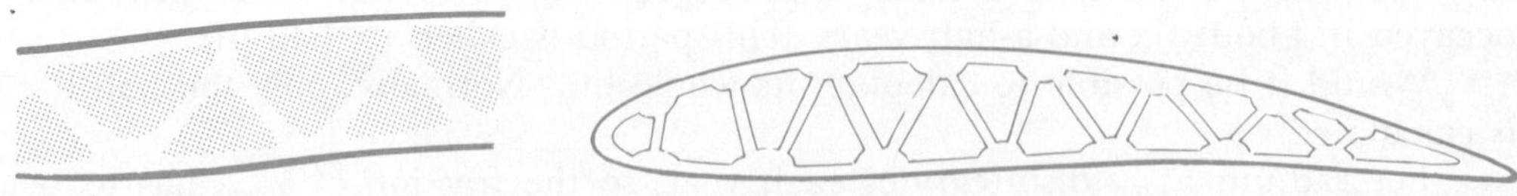

The bone is able to take stresses as well as it would if it were solid and the saving in weight is essential for flying.

CLASS DISCUSSION

What mathematical concepts, topics or shapes have been used by man in the construction of:

- a record
- a deck chair
- a ladder
- a soccer ball
- the Eiffel Tower
- organ pipes
- rope or a cable made of strands
- the Harbour bridge
- a revolving door
- a French horn
- a modern suspension bridge
- an astronomical telescope
- the hair-spring that drives the balance wheel of a clock.
- a guitar (with particular reference to the length of the strings)
- a crane
- a wheel
- a spiral staircase
- a hang-glider

Discuss the mathematics in:

a. planning a military campaign.
b. classifying library books.
c. a court-room trial.

DID YOU KNOW?

A fine example of projective geometry is the apparent sameness in size of the sun and the moon. The sun is 400 times larger in diameter than the moon, but it is also about 400 times further away from us, which is why the sun and moon *appear* equal in size to us. This is spectacularly illustrated whenever the moon comes between the earth and sun to produce a total solar eclipse. It appears that the disc of the moon almost exactly covers the sun's silhouette. During an eclipse, measurement of the sun's diameter and its distance from earth are carried out, using trigonometry.

On 23rd October 1976 a total solar eclipse blotted out the sun over Australia, New Zealand, New Guinea, Indonesia and most of Africa marking the fourth and final eclipse to be seen over Australia this century. It lasted only a few minutes, occurring about 4.30 p.m. Australians will not see another one until the year 2002.

Unit 10 Mathematics in Art

Art in Ancient Times

The ancient Egyptians were probably the first to use geometrical patterns extensively, although Neolithic man also developed a keen feeling for such patterns. The baking and colouring of pottery, the plaiting of rushes, the weaving of baskets and textiles and later the working of metals led to the cultivation of plane and spatial relationships. Dance patterns also played a role. Neolithic ornamentation used congruence, symmetry and similarity, with numerical relationships also playing a part. There were prehistoric patterns that represented triangular numbers and others with "sacred" numbers.

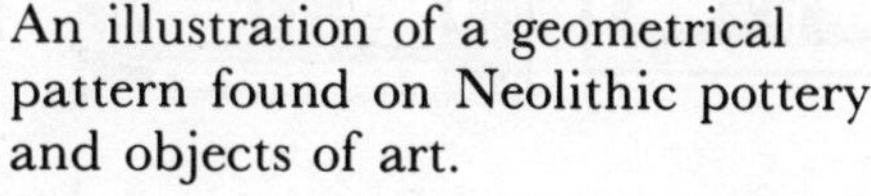

An illustration of a geometrical pattern found on Neolithic pottery and objects of art.

These three motifs are from urns found in Hungary and show attempts at the formation of triangular numbers, which played an important part in the Pythagorean mathematics of a later period.

Design from an Egyptian pot from the period 4000–3500 B.C.

ACTIVITY – ASSIGNMENT

1. Look up some information on the mathematical patterns found on vases of the Minoan and early Greek periods, on Byzantine and Arab mosaics, on Persian and Chinese tapestries. Draw illustrations of them, in colour, where possible.
2. Research the use of mathematics in the art of the Egyptians – their architecture as well as art.

Primitive Art

Even among very primitive tribes we find the use of simple, but very effective geometrical patterns. Naturally, the dominance of their environment, of the sun, moon and stars, necessities such as creeks and water holes, their "magic" and religious beliefs and many other factors are apparent in their cave drawings, on shields and implements, as well as on their pottery and dwellings.

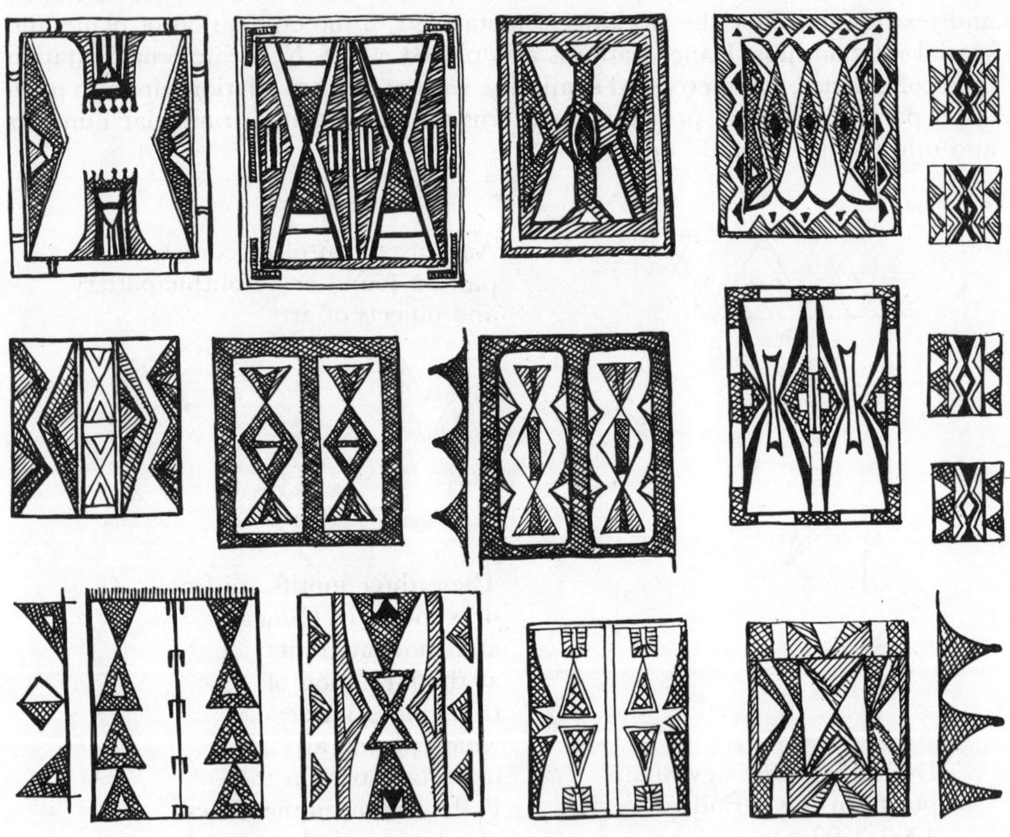

Geometrical patterns developed by American Indians (from Spier).

ASSIGNMENT

Choose a race of people and find out what you can about the mathematics–art relationship of that culture. Use illustrations.

The Renaissance – "Rebirth"

Projective geometry had its origins in the work of the Renaissance artists. Medieval painters had been content to express themselves in symbolic terms, with people and objects being treated in a highly stylised manner, often on a gold background. This seemed to emphasise that the subject of the painting, generally religious, had no connection with the real world. Look up Simone Martini's "The Annunciation".

With the Renaissance came not only the desire to paint realistically, but also a revival of the Greek doctrine that the essence of depicting nature is mathematical law, that is, depicting a three-dimensional world on a two-dimensional canvas. Leonardo da Vinci's "Last Supper" is a fine example.

Many artists since the Renaissance have been interested in geometrical paintings. One of the best-known was Piet Mondrian who died in 1944. His severe paintings, particularly those using rectangles, have had a dramatic impact on modern art. Look for his work entitled "Painting in Blue and Yellow".

Ben Nicholson is a more recent artist who has followed this same style.

A good work of art must have an aesthetic arrangement or composition. Many great artists have used the golden section, or the triangle, as the basis for their paintings or sculptures, in some cases with inverted or overlapping triangles being used. Memlinc's "Madonna and Child" or Rembrandt's "Portrait of a Soldier" are good illustrations of this technique. The inclusion of a triangle in a composition gives it a feeling of *stability*.

The inclusion of horizontal, vertical or curved lines also plays a part. Horizontal lines give a restful feeling, vertical lines give dignity and swirling curved lines give the impression of energy and movement. In many cases, two sorts of lines are used in the same composition.

What quality do you think is gained by using circles?

Another quality, noticeably lacking in pre-Renaissance works, was perspective (or space perception) the quality that gives a painting a three-dimensional depth and is derived from the Latin word for "seen through".

Cavemen did not use perspective. Instead, they demonstrated "randomness" (the decimal expansion of π is thought to be an example of randomness . . . 3·141592 . . .).

Historically, the first great exponent of perspective was Albrecht Dürer, the famous German artist who drew most objects realistically. This meant he had to introduce a certain degree of distortion, and so was forced to use mathematics.

Perspective

Projective geometry studies what happens to shapes when they are distorted in special ways.

Our eyes can deceive us!

Geometry tells us that parallel lines never meet, yet we seem to see railway tracks coming to a point in the distance. What do we call this optical phenomenon? **Perspective.**

By perspective we mean the change in dimension and direction the human eye produces in looking at solid objects. If two objects are the same size, the one farther from the viewer looks smaller.

Example 1: How the eye sees objects at different distances.

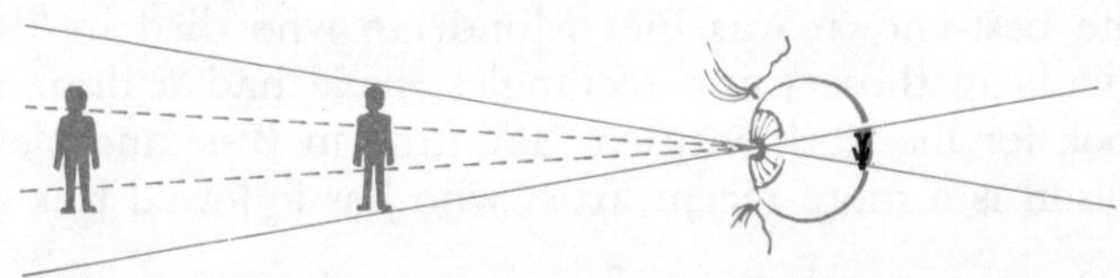

Actual size of the object as seen by the retina of the eye. The image **is upside down!**

The ratio of apparent decrease in dimensions states that as the distance of an object increases, its apparent size decreases in inverse proportion to the distance.

Example 2: A line of telegraph poles going almost directly away from the viewer.

Notice that the poles appear:

a. shorter,
b. thinner,
c. closer together as they converge.

This is because height, breadth and spacing all appear smaller as the distance increases.

There are two main kinds of perspective, parallel and angular. Parallel perspective is not a true type but is convenient to use when a surface is directly in front of the viewer.

Example:

A rectangular prism in which the front surface is shown as a rectangle.

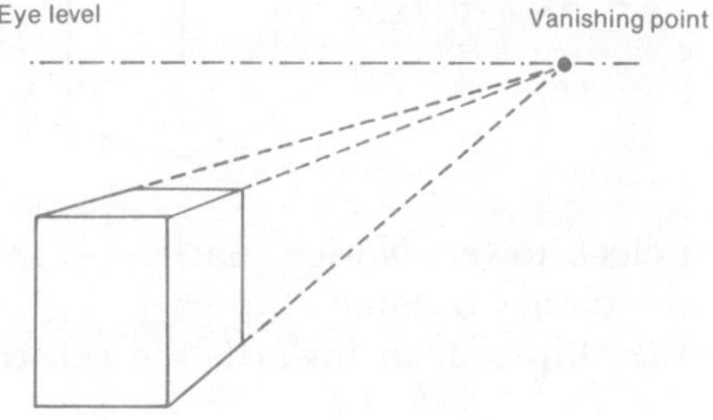

The same prism in angular perspective.

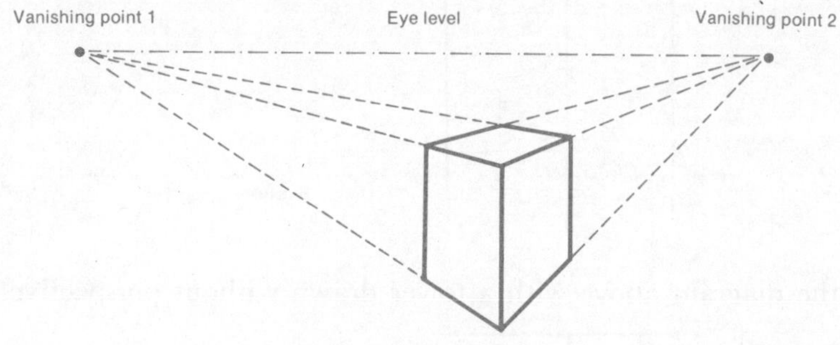

The eye level, as its name implies, is the plane extending from the viewer's eyes.

Example of eye level below prism.

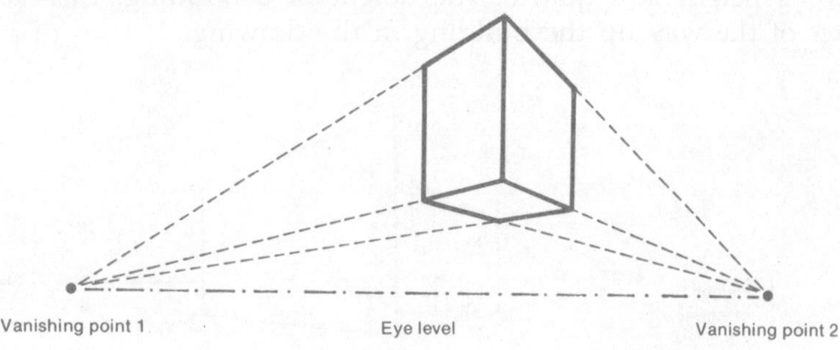

Example of eye level between top and bottom of prism.

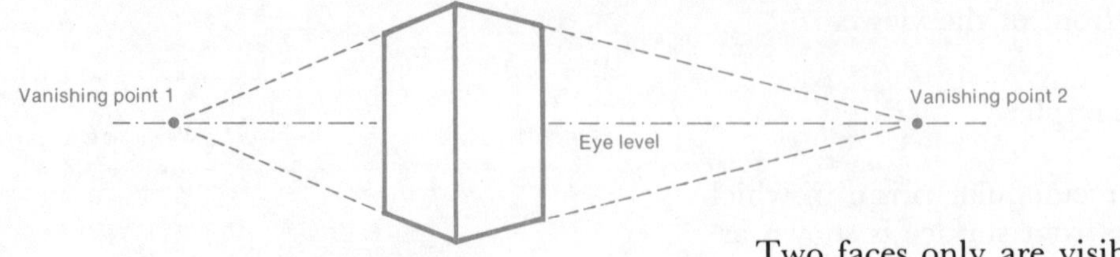

Two faces only are visible.

Here is a clock tower. Notice that:

a. the circles become ellipses,
b. the ellipses lean towards the centre.

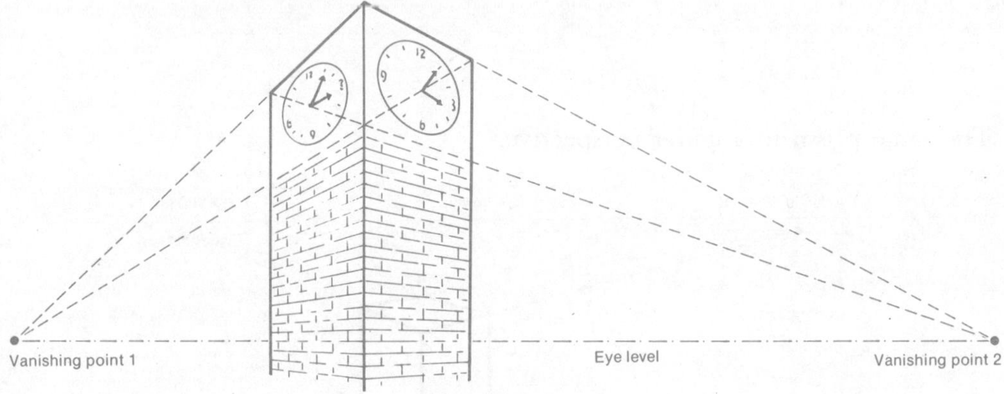

Compare the diagram above with a tower drawn without perspective.

If the viewer's height is a quarter the height of a building, the eye level will be a quarter of the way up the building in the drawing.

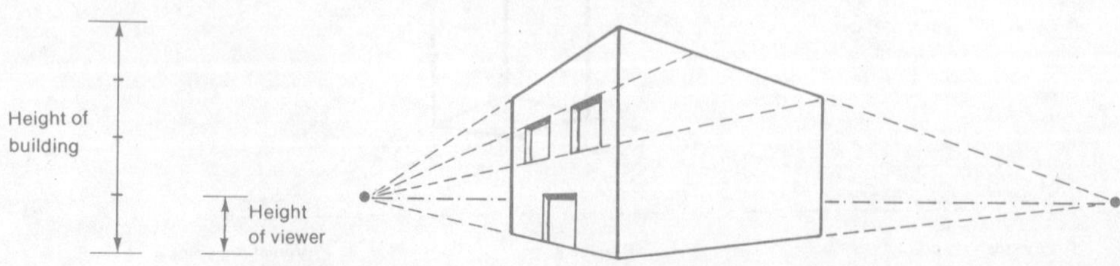

ACTIVITY 1

1. Observe a building at your school.
2. Compare your height with its height.
3. Draw the vertical edge nearest to you.
4. Draw in the eye level.
5. Choose two vanishing points, one closer to the vertical edge than the other.
6. Join the top and bottom of the edge to each vanishing point.
7. Draw in the other two visible edges of the building, using vertical lines.
8. Windows and doorways are then drawn in using the same method. (Draw the nearest vertical and join to the vanishing points first.)

ACTIVITY 2

To draw the interior of a room.

1. Draw one vertical corner line.
2. Draw in the position of your eye in relation to the line.

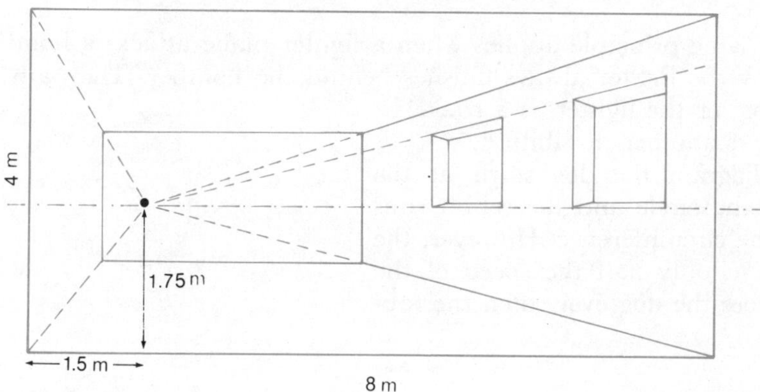

3. Mark in the other lines as shown.

Dürer wrote a treatise on perspective and even Leonardo da Vinci's notebooks are scattered with observations on the subject.

Curves of Pursuit

Suppose a rabbit, feeding in the middle of a field, sees a dog running directly towards it. The rabbit runs in a straight line to its hole while the dog runs at all times directly towards the rabbit. Diagrammatically this is the situation:

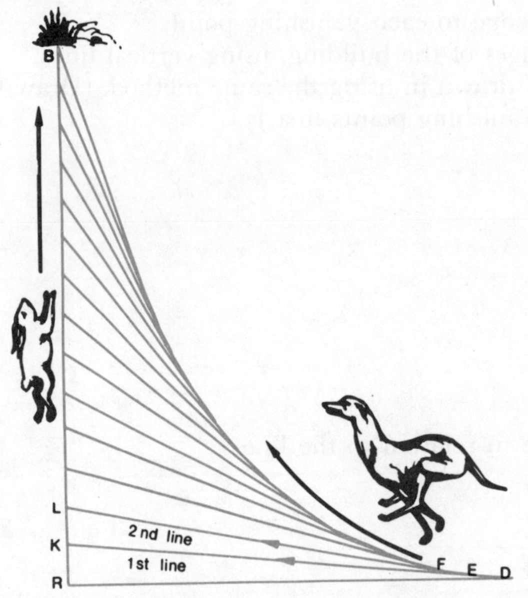

When the rabbit reaches point K it is seen by the dog which starts to head towards it along the first line. When the dog reaches E it notices that the rabbit has continued to run and is then at point L, so it changes course and heads towards L, along line 2. The process is repeated over and over again.

RK = KL = DE = EF (dog and rabbit each covering equal distances in the same time).

This same principle applies when a fighter plane attacks a bomber and explains why the fighter always finishes behind the bomber. The path traced out by the dog, or the fighter, is a **tractrix**.

Here is another possibility:

In this diagram the dog starts at the centre of the circle and the rabbit runs around the circumference. However, the dog runs at only half the speed of the rabbit. Does the dog ever catch the rabbit?

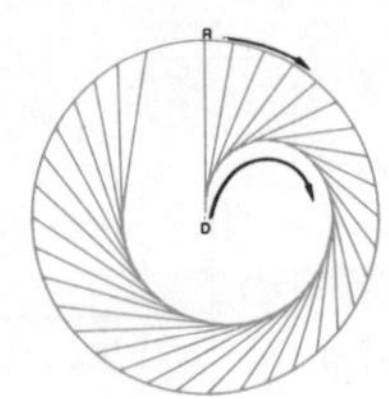

ACTIVITIES

- Draw the first example on a large sheet of paper making the vertical line as long as possible.
- For a variation, try having the rabbit run around the perimeter of a square field and start the dog either at the centre, or at the corner opposite the rabbit.

Vary the speed so that sometimes the rabbit runs faster and sometimes slower than the dog.

- If four mice start in the four corners of a square room, then move at a constant speed towards their moving objective, how far do they move in the spiral?

If three elements are involved in curves of pursuit, each moving directly towards the next in an anti-clockwise direction, the elements would eventually move in a circle (when the force had reached its maximum value) as the force required to produce the acceleration to maintain the path would not be possible.

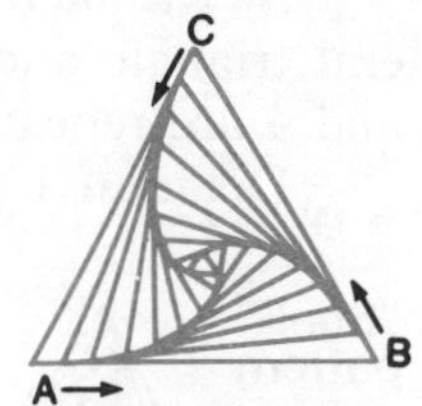

Diagram 1

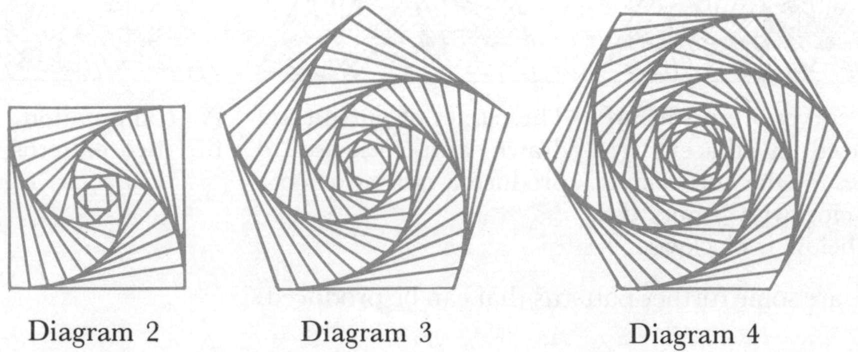
Diagram 2 Diagram 3 Diagram 4

These diagrams illustrate what happens when 4, 5 and 6 elements are involved.

Can you see the rotation effect of the square, pentagon and hexagon when you look at the centre of each diagram?

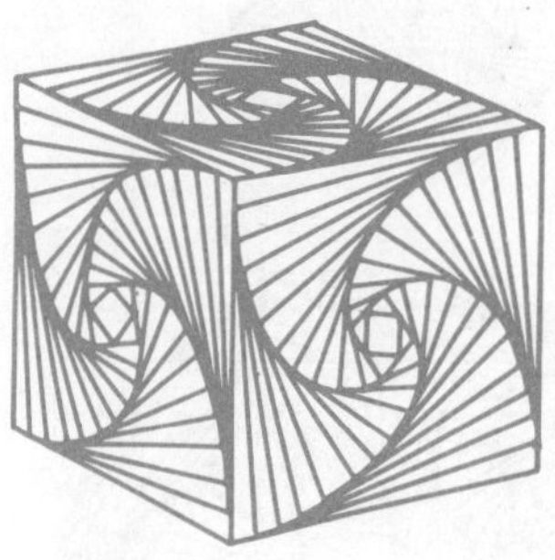

A three dimensional design based on the four mice problem.

ACTIVITY

1. Draw an 80 mm square and mark points on its sides at equal distances (say 10 mm) from the four vertices. Join them, to form a similar square, then repeat the process as many times as you can.

2. Do the same for a pentagon and hexagon.

As was mentioned earlier, only the regular polygons, the square, the equilateral triangle and the hexagon will tessellate. This means that if diagrams 1, 2 and 4 are repeated and fitted together, they will form interesting effects.

Diagram 1 can be used to form these patterns:

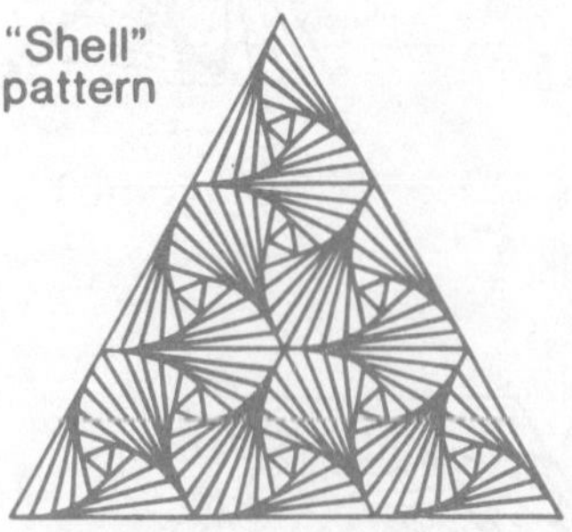

The curves of pursuit are reversed in adjacent triangles – the top one is anti-clockwise while the one below is clockwise.

The curves of pursuit are drawn anti-clockwise to produce a twisted effect.

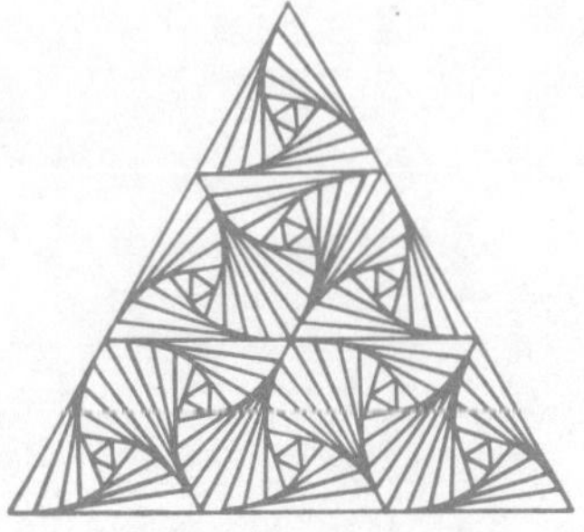

A combination of the first two patterns.

Here are some further patterns that can be produced.

Modern Art

Today, the science of art cannot be separated from the art of science. The elements of time, motion, space, optics and perception are the concerns of both the artist and scientist. New interpretations and expressions arise from our advancing scientific age and are reflected in many optical or perceptual abstractions.

What is Amiss?

Study this engraving by England's William Hogarth (1697–1764) in which he has made deliberate mistakes in the clues of space relationship. For example the sheep appear larger as they get further away! What is wrong with the barrels? buildings? hiker? Make a list of all the mistakes, in your book.

"False Perspective"
by William Hogarth.

M. C. Escher also incorporated "impossible figures" (a play with laws of perspective) and optical illusions in his works, with most unusual treatment of planes and dimensions. Study these.

Relativity, 1953, by M. C. Escher.

Waterfall, 1961, by M. C. Escher.

In the lithograph "Belvedere" an "impossible" cube is sketched on a piece of paper lying at the foot of a seated boy, who is holding a skeletal model of this unrealisable figure! A youth near the top of a ladder is outside the belvedere while the base of the ladder is inside . . .

Belvedere, 1958, by M. C. Escher.

Another fascinating lithograph "Ascending and Descending" derives from a perplexing impossible figure of monks marching around a stairway on the roof of a monastery, the outside monks climbing, the inside monks descending, endlessly . . .

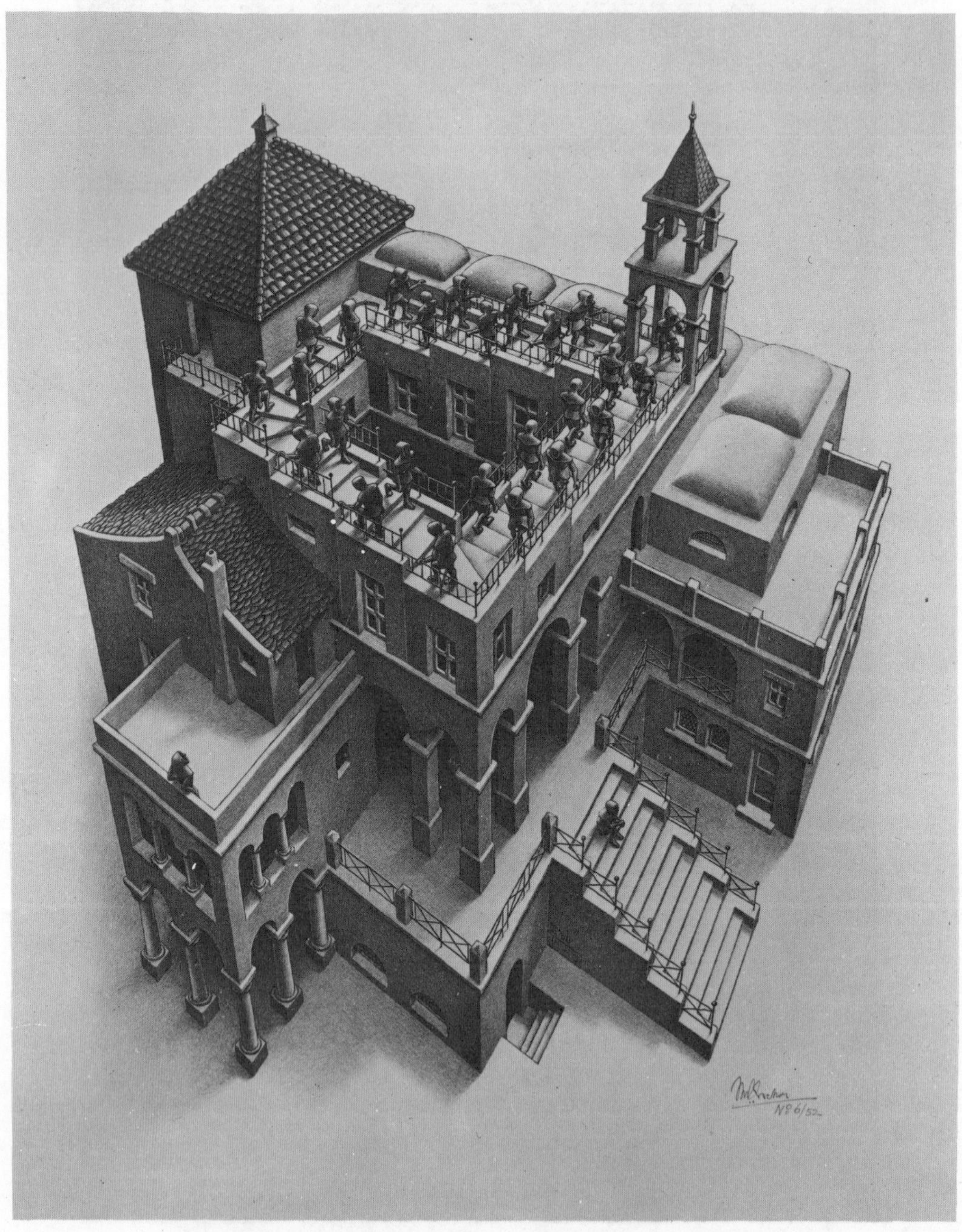

Ascending and Descending, 1960, by M. C. Escher.

Optical (Op) Art

This art has been popular since the mid 1960s. It is the name for strictly mathematical "abstracts" with a great emphasis on perception. There is no understanding needed, for the response is direct. Even though mathematical analysis may be applied to optical art, it must be remembered that the artist does not always approach his art with much research in this field. The Mayans, Moors, Orientals and Egyptians used optical effects. Look up some books on optical art or find some illustrations of the works of Bridget Riley and Victor Vasarély. Sometimes the effects are intended to dazzle the eye – vivid colours, optical illusions, striped and dotted patterns that play tricks on reality.

1.

Roman tile design.

Which tessellating pentomino can be seen here?

2.

Japanese fret.

3.

Greek fret.

Optical Progression.

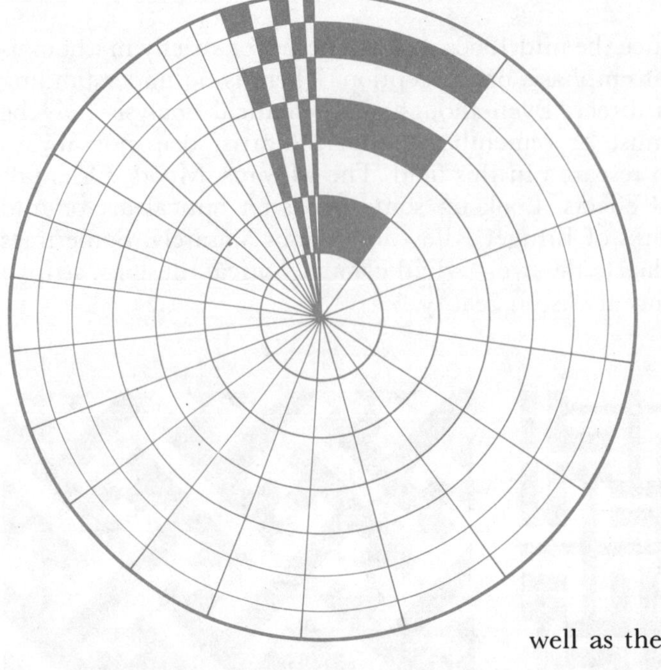

Use your instruments to draw a larger diagram of this. You could make the concentric circles different widths as well as the radii. Continue colouring.

Construct your versions of these two designs in your exercise book.

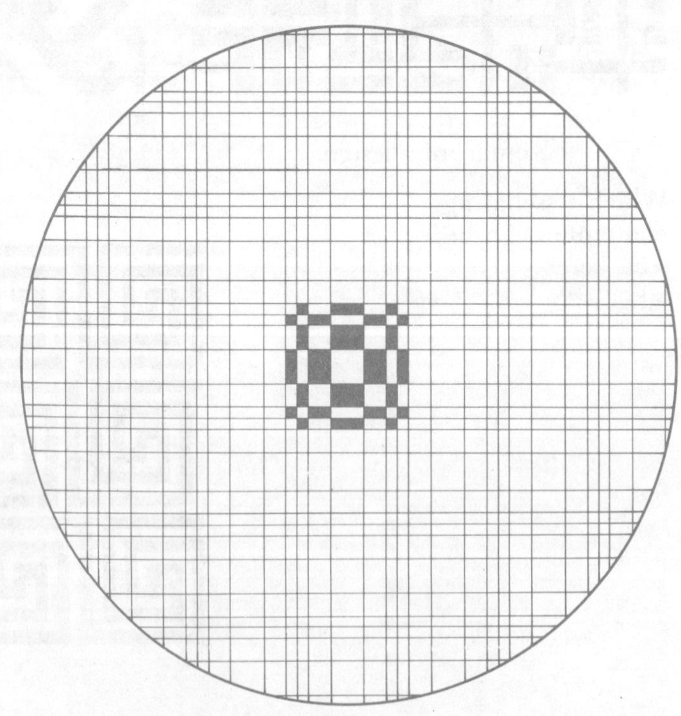

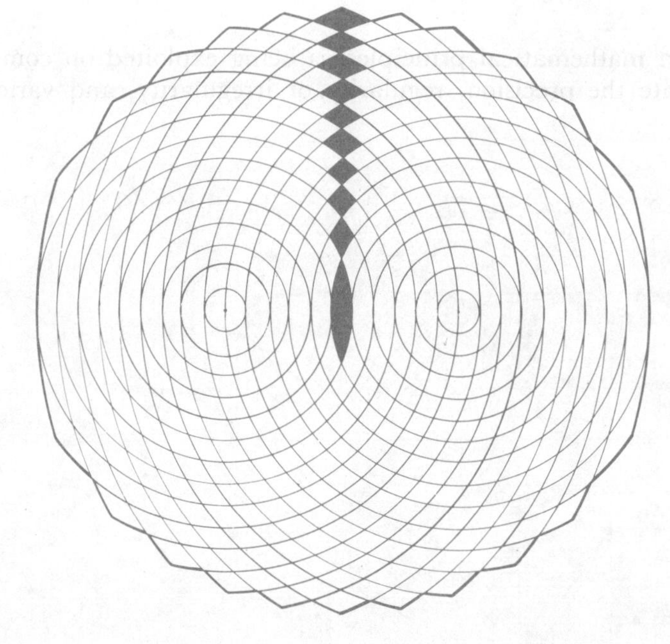

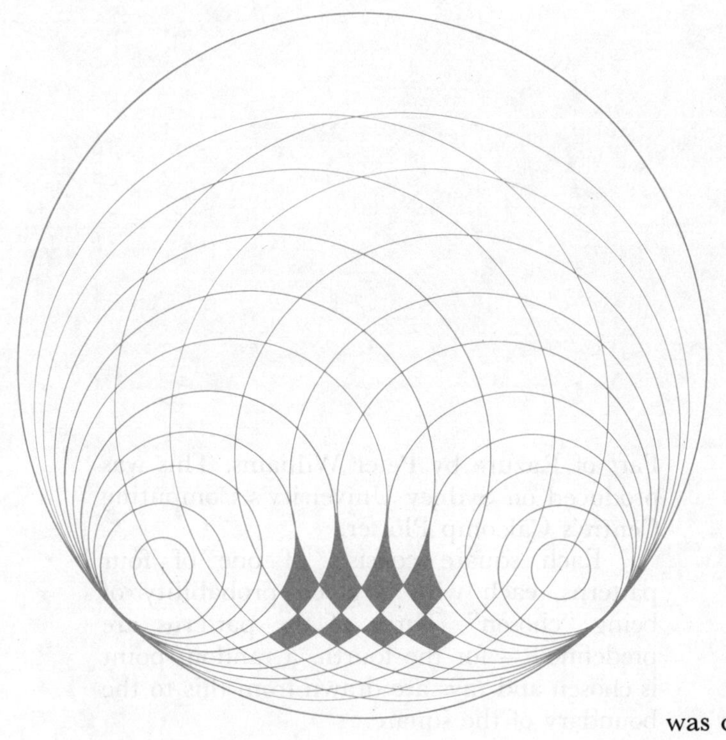

This is a variation of the previous design.

Can you see how it was constructed?

Computer Art

Art, based on mathematical principles, is being exploited on computers. These works illustrate the precision, regularity or irregularity, and variety of designs possible.

Part of Razure by Peter Williams. This was produced on Sydney University's Computing Centre's Calcomp Plotter.

Each square consists of one of four patterns, each with a given probability of being "chosen". Three of the patterns are predefined – for the fourth, a random point is chosen and rays are drawn from this to the boundary of the square.

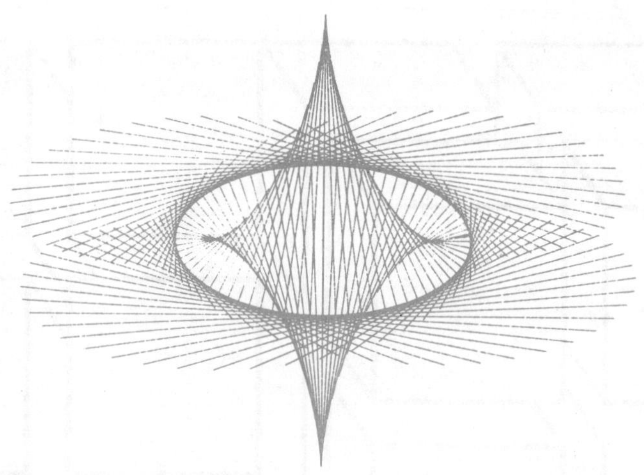

Humming bird produced on a Calcomp 960 Plotter.

Part of Leaves by Joan Wilcox.

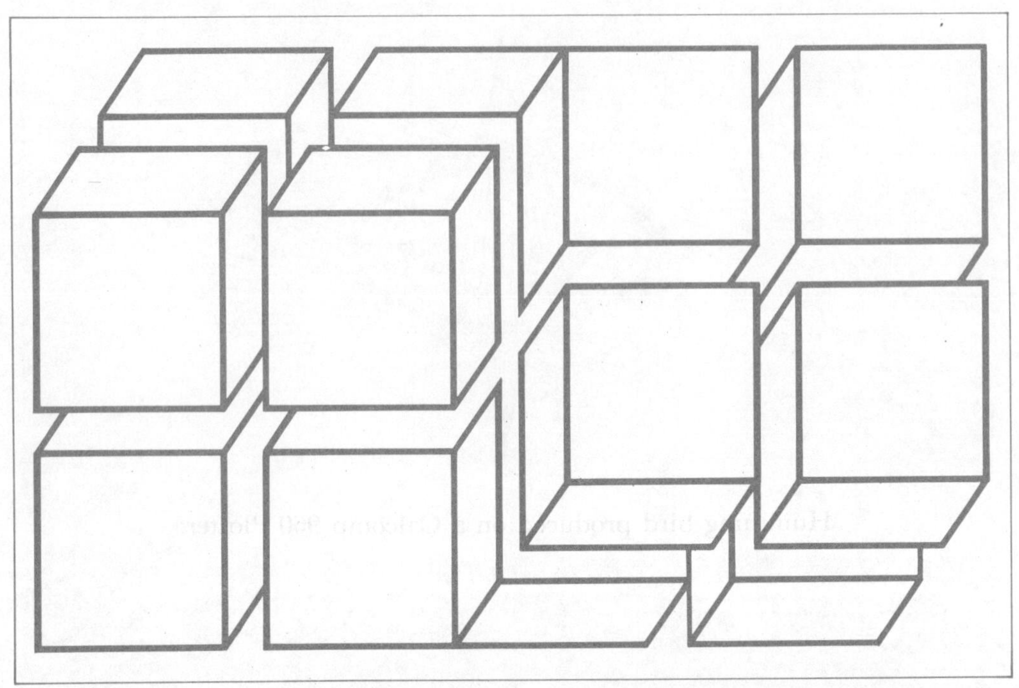

An example of an impossible perspective drawing drawn by a computer.

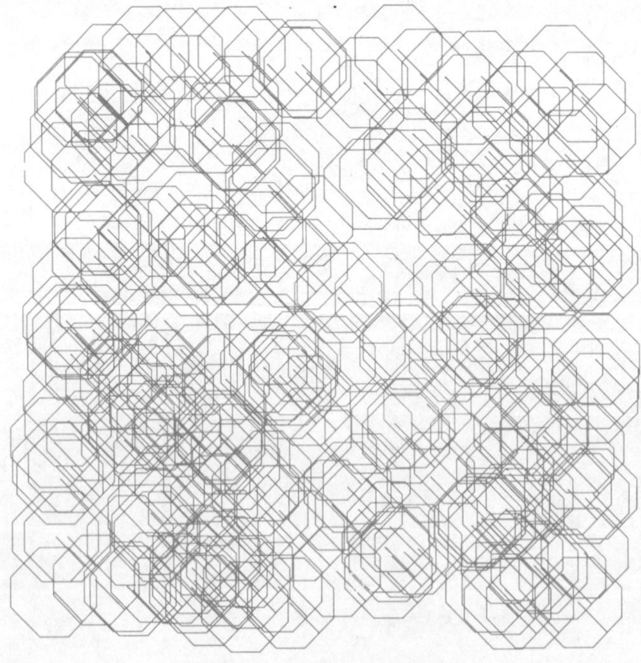

Space by Dao Khanh.

Spiral produced on a Calcomp 836 Plotter.

Indian blanket by Bruce McNair.

Solutions

Unit 1 A Hint of Magic

Magic Hexagons

It is impossible to complete an order 2 magic hexagon. If the digits 1 to 7 are added and the total divided by 3, the resulting 28 is *not* evenly divisible by the 3. There is no magic constant as by definition, it must be an integer (whole number).

Magic Word Squares

P	E	T	A	L
E	R	A	S	E
T	A	S	T	E
A	S	T	E	R
L	E	E	R	S

OR

S	E	P	A	L
E	R	A	S	E
P	A	S	T	E
A	S	T	E	R
L	E	E	R	S

Unit 2 Topology

Lewis Carroll's Network Problem

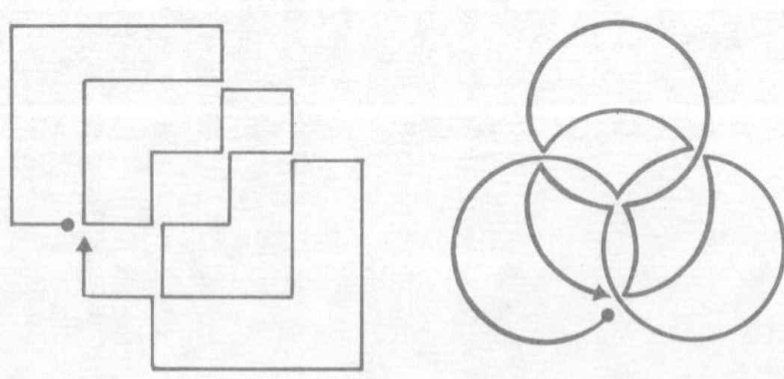

Network Problem

The route is A–C–B–D–E–F–G–I–J–H–K (72).
The instructions lead to room 13 via 5–6–10–13.

Belts

Yes. C and D go clockwise while B goes anti-clockwise. The wheels will turn if all four belts are crossed, but not if one or three belts are.

Teasers

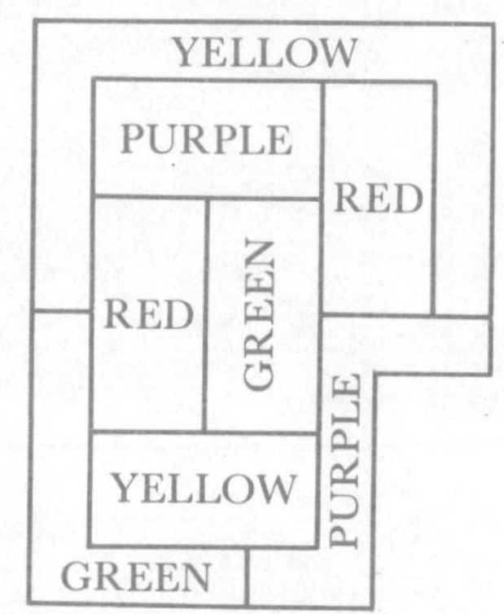

- 3 colours
- 5 colours
- 8 m^2 of red is mixed with 8 m^2 of blue to make 16 m^2 of purple

Topological Tomfoolery

1. It cannot be done.
3. Fold your arms first, pick up the end carefully, then unfold your arms.
4. Push buttonhole through the loop, then over the end of the pencil.

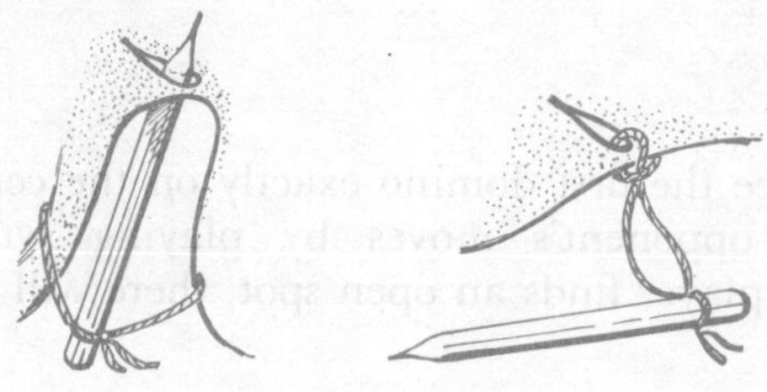

Unit 3 Pastimes

Billiards

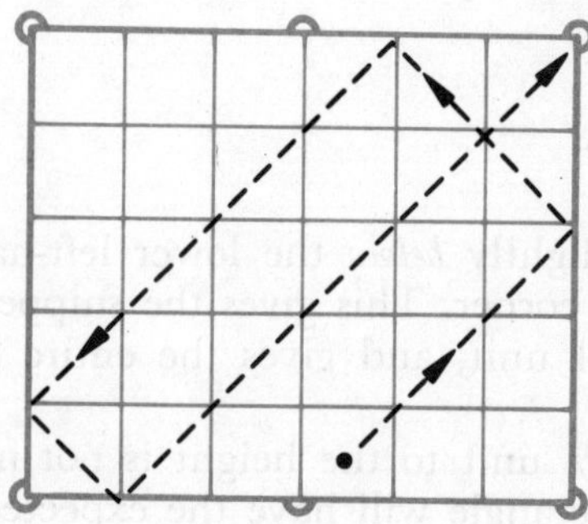

Dominoes – Square

Top: [4|0] [4|3] [3|3] [3|1] [1|1] [1|4] [4|6] [6|0]

Left (top to bottom): [4|0] [0|0] [0|1] [1|6] [6|6] [6|5] [5|2]

Right (top to bottom): [6|0] [0|2] [2|4] [4|4] [4|5] [5|5] [5|1] [1|2]

Bottom: [2|2] [2|6] [6|3] [3|0] [0|5] [5|3] [3|2] [2|1]

Magic Square

4	0	3
6	3	5
0	2	3
5	5	6
0	5	2
6	6	2

Symmetry Strategy

The strategy is to place the first domino exactly on the centre of the board and then to match the opponent's moves by playing symmetrically opposite. Whenever the second player finds an open spot, there will always be one to pair with it!

Chess

64.

No. A knight's move takes it to a square of a different colour. Therefore, on a 5 x 5 board, there must be thirteen of one colour and twelve of the other.

Thirteen knights cannot leap to twelve squares without two of them landing on the same square.

Chessboard Paradox

The oblique line passes slightly *below* the lower left-hand corner of the square and the upper right-hand corner. This gives the snipped off triangle an altitude of $1\frac{1}{7}$ units, rather than 1 unit, and gives the entire rectangle a height of $9\frac{1}{7}$ units.

The addition of the $\frac{1}{7}$ unit to the height is not noticeable, but when it is taken into account, the rectangle will have the expected area of 64 units^2.

Knight's Areas

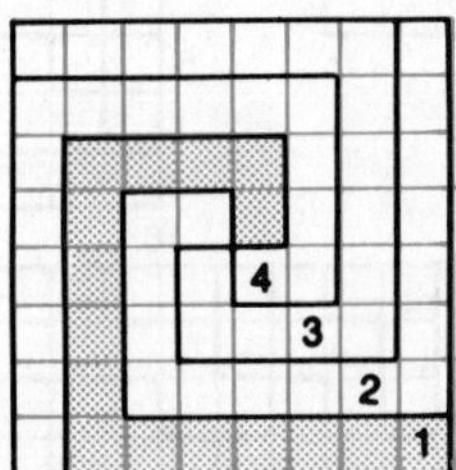

Seventeen Lines

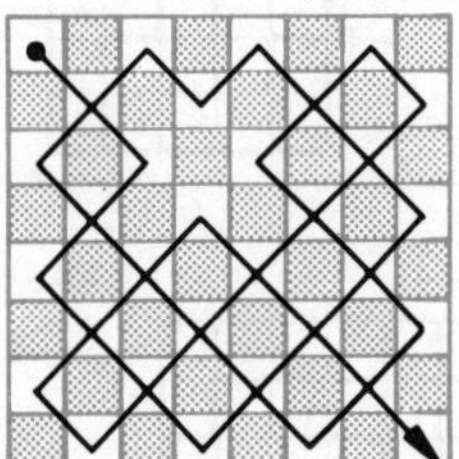

Unit 6 Polyominoes

Tetromino Activities

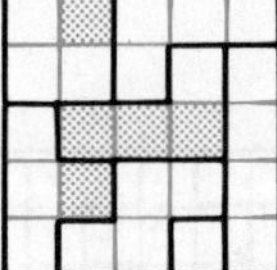

Pentominoes

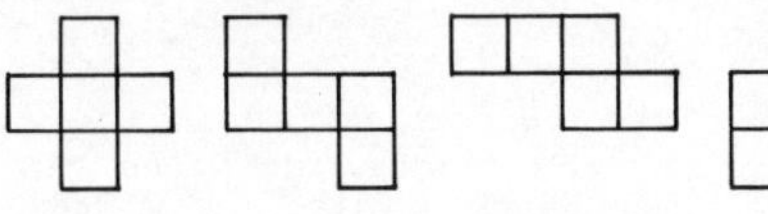
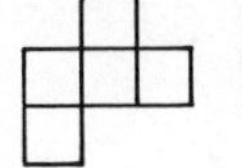
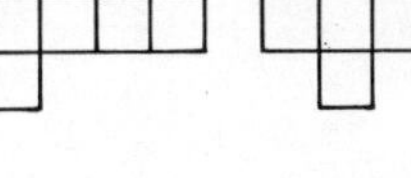
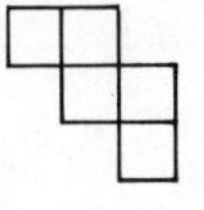

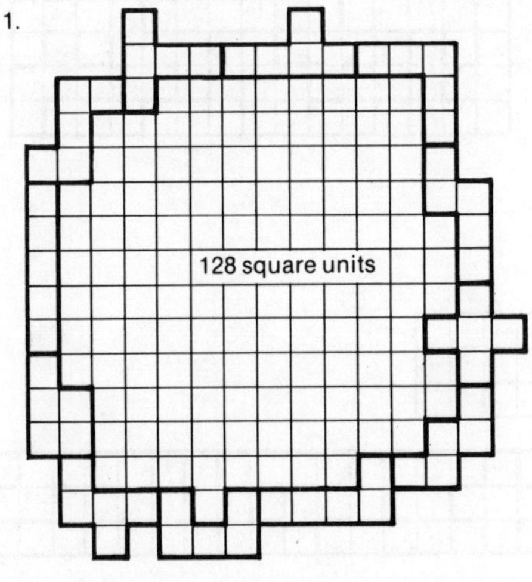

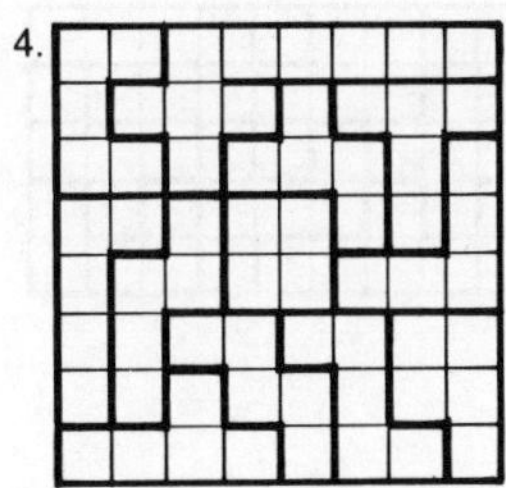

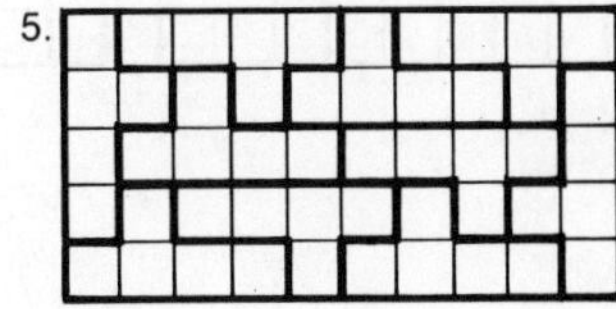

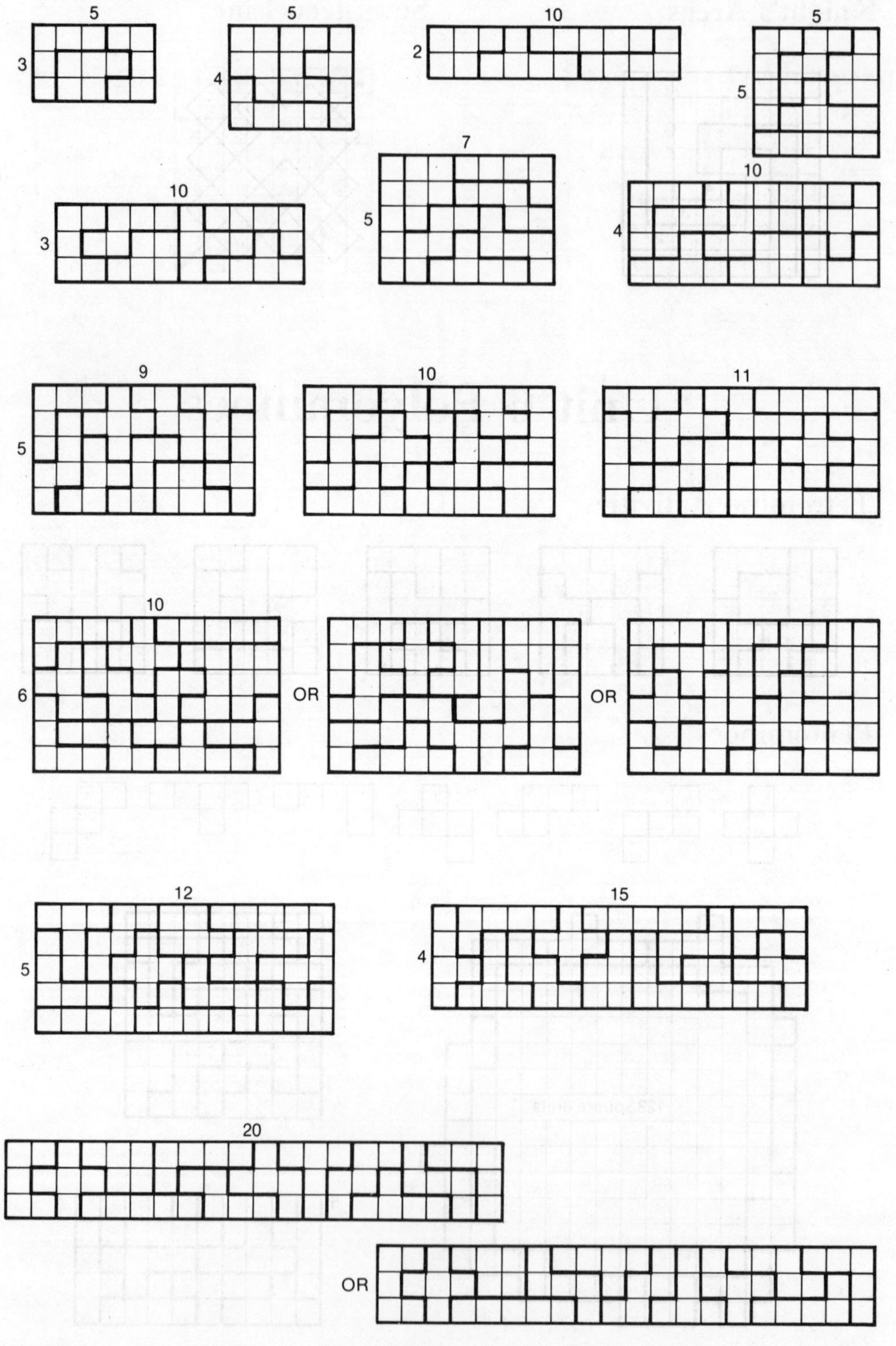
5
3
5
4
10
2
5
5
7
5
10
3
10
4
9
5
10
11
10
6
OR
OR
12
5
15
4
20
3
OR

Duplicates

6.

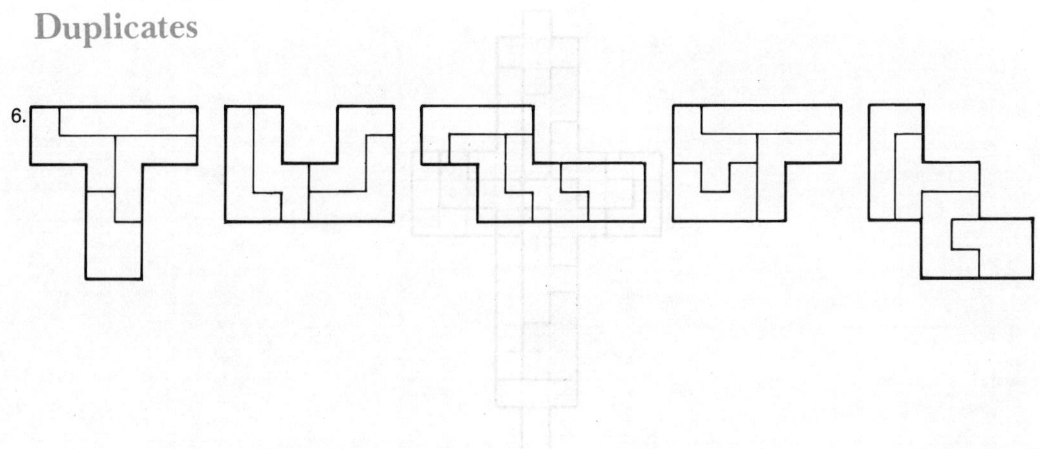

Pentominoes

8.

There is only one crossroad.

Domino Game

No, you cannot cover the board because there are an unequal number of black and white squares left.

Polyhexes

The **triangle** is impossible.

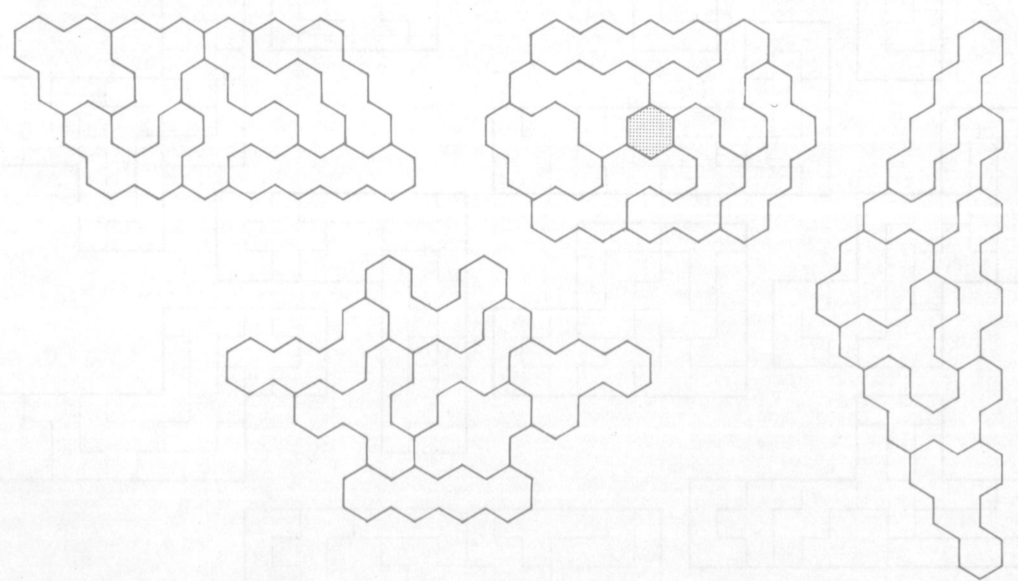

Pentahexes

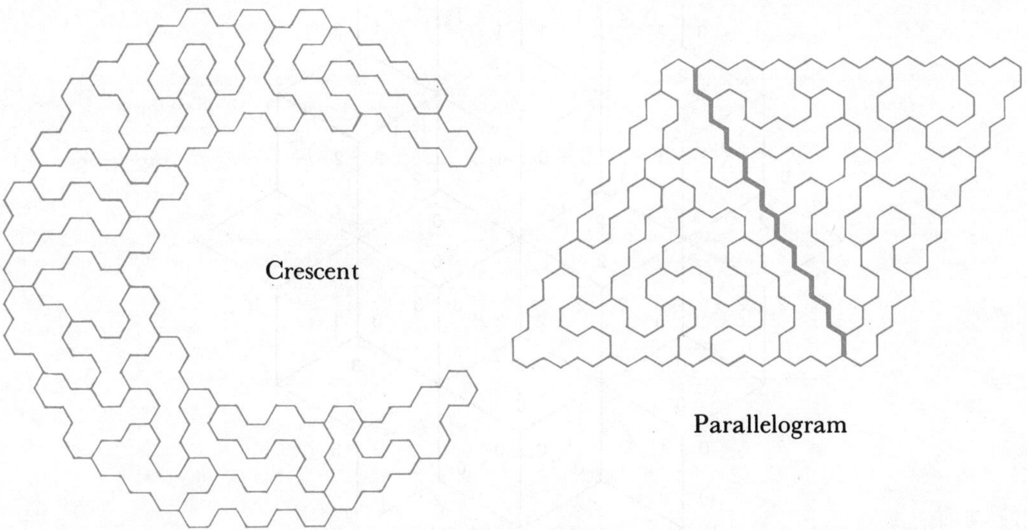

Polyiamonds

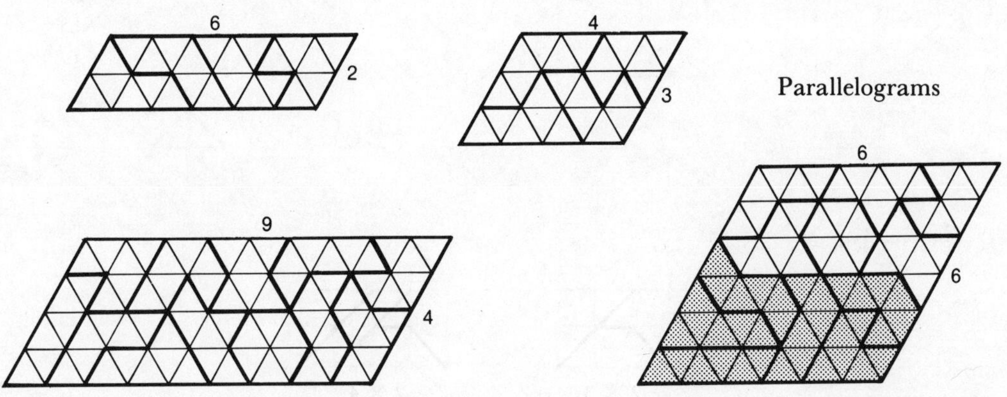

Rhombus: Note the division of the figure into two congruent halves.

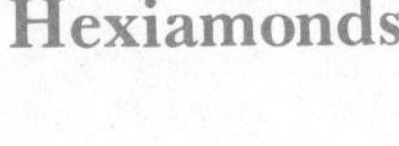

Hexiamonds

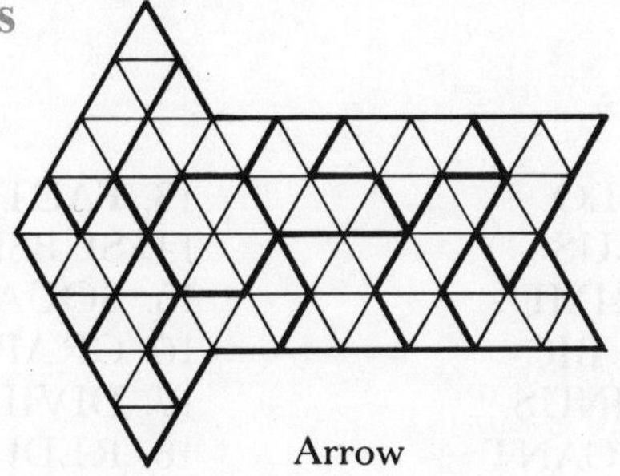

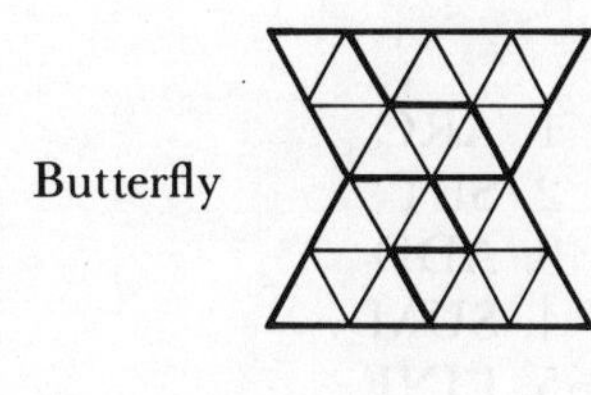

Tromino Puzzle

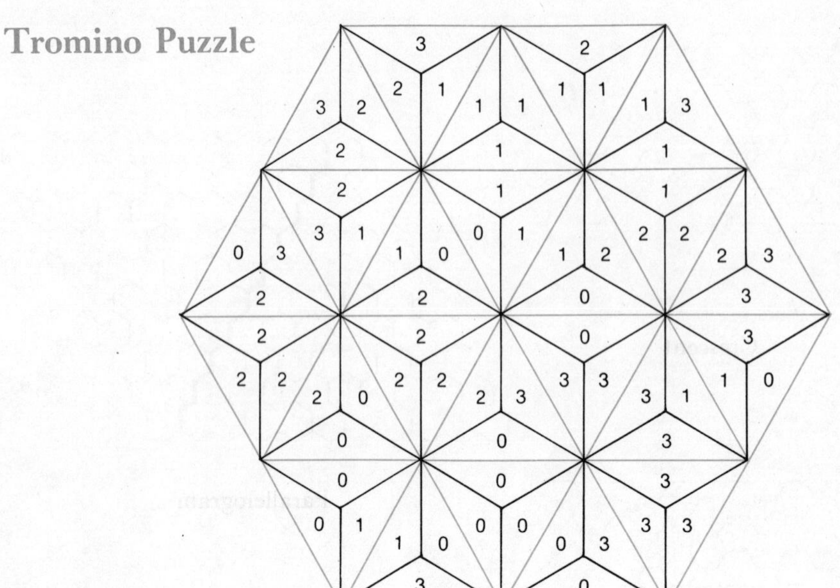

Polyaboloes – Fourteen tetraboloes

Rectangles

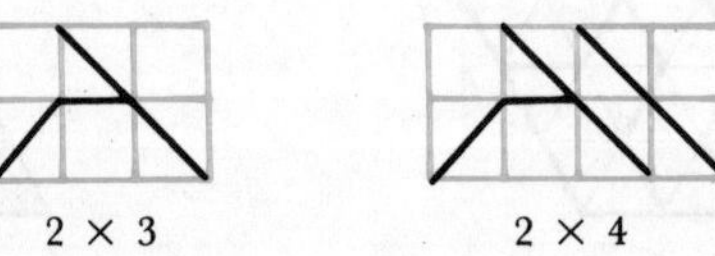

Unit 8 Number Relationships

Cryptics

1. ARC
2. SET
3. ADD
4. SUM
5. LINE
6. AREA
7. KILO
8. PLUS
9. PRIME
10. CUBIC
11. MINUS
12. SECANT
13. FACTOR
14. SUBSET
15. SQUARE
16. GRAPHS
17. DIVIDE
18. REDUCE

Index